SPATIAL CONTEXT: AN INTRODUCTION TO FUNDAMENTAL
COMPUTER ALGORITHMS FOR SPATIAL ANALYSIS

International Society for Photogrammetry and Remote Sensing (ISPRS) Book Series

ISSN: 1572-3348

Book Series Editor

Zhilin Li
Department of Land Surveying and Geo-Informatics
The Hong King Polytechnic University
Hong Kong, P.R. China

information from imagery

Spatial Context: An Introduction to Fundamental Computer Algorithms for Spatial Analysis

Christopher M. Gold

Southwest Jiaotong University, Chengdu, China and LIESMARS, Wuhan University, Wuhan, China

CRC Press
Taylor & Francis Group
Boca Raton London New York

CRC Press is an imprint of the
Taylor & Francis Group, an **informa** business

A BALKEMA BOOK

CRC Press
Taylor & Francis Group
6000 Broken Sound Parkway NW, Suite 300
Boca Raton, FL 33487-2742

First issued in paperback 2019

© 2016 ISPRS
CRC Press is an imprint of Taylor & Francis Group, an Informa business

Typeset by MPS Limited, Chennai, India

ISBN-13: 978-1-138-02963-7 (hbk)
ISBN-13: 978-0-367-87892-4 (pbk)

Library of Congress Cataloging-in-Publication Data

Names: Gold, Chris, 1944– editor.
Title: Spatial context : an introduction to fundamental computer algorithms
    for spatial analysis / editor, Christopher M. Gold, University of South
    Wales, Cardiff, UK.
Description: London, UK : CRC Press/Balkema is an imprint of the Taylor &
    Francis Group, an Informa business, [2016] | Series: International Society
    For Photogrammetry and Remote Sensing (ISPRS) book series ; 8 | Includes
    bibliographical references and index.
Identifiers: LCCN 2016010598 (print) | LCCN 2016016489 (ebook) | ISBN
    9781138029637 (hardcover : alk. paper) | ISBN 9781498779104 (eBooks PDF) |
    ISBN 9781498779104 (ebook)
Subjects: LCSH: Geospatial data. | Geographic information systems—Mathematics. |
    Computer algorithms. | Algebras, Linear.
Classification: LCC G70.217.G46 S63 2016 (print) | LCC G70.217.G46 (ebook) |
    DDC 910.285—dc23
LC record available at https://lccn.loc.gov/2016010598

Visit the Taylor & Francis Web site at
http://www.taylorandfrancis.com

and the CRC Press Web site at
http://www.crcpress.com

Table of contents

Preface

The purpose of this book is, to borrow an old British Army expression, 'square-bashing'. My children grew up knowing that squares were bad and triangles were good: I hope I can persuade you, the reader, likewise!

This book is the result of over 30 years of puzzling over the problem of Space. Great scientists have done this before – Newton, Leibnitz and Einstein among others. We cannot copy them. However, more recently the problem has become particularly acute with the arrival of digital computers, finite-precision arithmetic and, in my particular discipline, the desire to store geographic information in the computer.

Clearly many people have done so, and successfully. But if you look more closely at the methods used they seem to be a hodgepodge of desperately-selected measures to resolve the same classes of problems, over and over again – whether in GIS, CAD or any other spatial discipline. This book is deliberately intended to go against the grain: something is not necessarily good just because it works (approximately) on faster and faster machines with more and more storage. It is good if it leads to a clearer understanding of how to phrase a new question, in a form that can be resolved using a fairly simple toolkit. Complexity of methods may often be necessary, but it is not always necessary.

Similarly, a method is not necessarily good just because it uses concepts we are comfortable with, ever since elementary school: often these were over-simplifications, or not sufficiently general. On the other hand, our intuition is often pretty smart: if we can see the answer (perhaps of the spatial relationships between a set of objects) and the computer cannot – then perhaps we can rephrase the methods to look more like ours, and less like a computer solution – until we run into the brick wall of a digital computer's limitations, and have to back-track.

Having worked in geology, geography, forestry, surveying, agriculture, water resources and computer science, I have tried to filter out a variety of concepts of space that we have used in common – trying to see where we are using clumsy, awkward, special-purpose techniques in a particular discipline, when we could be using more general and cross-disciplinary techniques to better advantage. We must bear in mind two depressing adages: almost nobody reads papers and textbooks from other people's disciplines (and many textbooks simply replicate the old-fashioned methods of their predecessors); and a general-purpose algorithm cannot, by definition, be faster or better than the best special-purpose one. Nevertheless I believe it is time to clean up as much as possible, in the hope that at least some of the concepts may filter into the next generation of textbooks, for students as yet unborn.

In this book I have attempted to clear out the underbrush around many ways of handling spatial objects, until there is a reasonable chance of the visitor seeing the larger view. I have not discussed, except in the most general terms, the attributes (colour etc.) attached to those objects – many other publications do this, and to me the hard problems come first, in the spatial relationships between them. In brief: the Dual is the Context.

Acknowledgements

A book – any book – is a collaboration between the author, his people, and the world around him: no man is an island. This book is no exception, so the question is: does he thank the many, or the few? I will start in the centre and move outwards.

The late great Victor Borge once started a concert by thanking his mother for making it possible, and his son for making it necessary: this is a plan. Nothing would have been possible without my parents struggling to make ends meet, and my consequent acceptance at a good boarding school in England when my father died. Much of my gratitude here must remain unexpressed. Hard work and a lot of letter writing got me a scholarship to Canada, where I obtained my BSc, MSc and PhD – this last only possible with the devotion of my wife Valerie: again so much must remain unexpressed here. My academic career demanded several moves – not always easy on growing children: my apologies and love to Karen, Pamela, Derek and Tim – and of course to Valerie. I pray that they feel at least partly recompensed.

My students and colleagues made this book necessary – at least, they demanded that I write it, and I backed off, saying it would be too much work: a sentiment with which most authors would agree, and which I, in retirement and with no more excuses, confirmed. Finally, my back to the wall, I promised Chen Jun and Li Zhilin that I would write it before the end of Chen Jun's presidency of the ISPRS – and the Publishers very kindly confirm that this will be possible. My thanks to both of them for the comradeship and arm-twisting, to Mike Shamos for explaining Voronoi and Delaunay concepts to a beginner, to Jack Snoeyink for introducing me to Computational Geometry, to Li Deren for encouraging my work on Voronoi diagrams, to Liu Guoxiang for support in recent years, to Michael Goodchild for encouragement since the early days and to Rien van de Weijgaert for showing me the structure of the Universe.

One of the major joys of the academic life is the excitement of developing new ideas with one's research students: I hope they felt the same. My students from Laval University, Quebec; The Hong Kong Polytechnic University; and The University of South Wales (was: Glamorgan) were: Weiping Yang, Mir Abolfazl Mostafavi, François Anton and Darka Mioc from Laval University; Krzysiek Matuk from Hong Kong; Hugo Ledoux and Rebecca Tse from both Hong Kong and South Wales; and Pawel Boguslawski, Rafal Goralski, Maciej Dakowicz and Neil Sang from South Wales. Their work is cited in these pages. Their friendship and diligence remains with me.

Research Chairs from NSERC (Canada) and the European Union provided the resources. I hope they consider it time and money well spent.

And, above all, my gratitude to God for giving us such a fascinating Universe to explore.

Benedicamus Domino!

Spatial Context: An Introduction to Fundamental Computer Algorithms for Spatial Analysis – Gold
© 2016 Taylor & Francis Group, London, ISBN 978-1-138-02963-7

Introduction

As this book is intended for a wide (read: undefined) audience I will start by giving a simplified overview of some basic mathematical ideas that are trivial to the expert but often overlooked by, or not described clearly enough to, students in some applied disciplines. I work on the assumption that they, like me, are more comfortable with geometric explanations than algebraic ones: the mathematician Roger Penrose admitted the same in one of his books, so we are in good company.

After a brief puzzling over all of this I conclude that coordinate systems – and especially fixed-precision ones inside computers – are a large part of our problem. The first section will give a simple-minded view of vector algebra, from a geometric point of view. Hopefully it will show that grasping a few simple (don't they all say that?) concepts is worthwhile, to push those pesky coordinates a little further away from the problem they are trying to confuse.

The second section attempts the same feat for matrix algebra. The knowledgeable will say: "Aha! Simultaneous equations!" – but they will be wrong. The simple determinant, expressed in 2D or 3D, auto-magically resolves many problems that you did not expect to be able to handle: finding if a line segment intersects a triangle in 3D space, for example. It is easy when you know how. Simple matrix multiplication is also behind all modern computer graphics: the concatenation (stringing together) of a variety of separate movements of an object allows us to track it very efficiently in our computer games. We then look at something computers really are good at: handling graphs, those collections of points and connecting edges that make up most of our final maps.

We then start to think about our real objective – things in space. Initially, and for quite a while, we will restrict ourselves to points: how can I tell the computer about the relative positions of my data points in two-dimensional space? This brings us into the realm of computational geometry, in most cases a fairly advanced discipline concerned with rigorous mathematical proofs. Luckily, our own particular needs focus on one particular structure – the Voronoi diagram – and we can use their results without having to be too concerned about proving the properties that we state.

Briefly, our problems with managing, and thinking about, space are twofold: firstly we think about space in different ways depending on the problem we are addressing; and secondly a computer doesn't think about space at all, but depends on the manipulation of entirely mythical sets of coordinates to arrive (with difficulty) at an approximation of what we, with brains trained for millions of years for the task, see to be obvious.

But before all that (but after the maths revision) I want to take you on a real journey into space – to look at the Cosmos, and how it might be structured, and what it may tell us about managing space at a scale we can do something about – a space consisting of voids and boundaries.

This gives us a framework for looking at a wide variety of cartographic processes. For independent points we will look at individual Voronoi cells, and at the dual Delaunay triangulations – as well as at simple point insertion, and dynamic point deletion, leading to effective surface interpolation methods and kinetic point movement algorithms and applications. Boundaries may be formed by lines of points, clusters, double lines of points or by solid segments composed of half-line pairs. Applications, including crusts and skeletons, contour and watershed handling, scanned map processing and solid line-work for urban mapping, are then readily classified and the algorithms and data structures described: this leads to the Unified Spatial Model for 2D mapping.

The last section takes us to 3D space – both simple shell models, e.g. for building exteriors, and true 3D, where volumetric elements may be defined, such as rooms and corridors. Shell models require some introduction to basic computer-aided design structures, and solid 3D requires advanced edge-based data structures – but both of these application areas lead to straightforward modelling techniques and emphasize the moral of this book: the spatial dual is the context.

Additional resources may be found at CONTEXT.VORONOI.COM.

Chapter 1

Preliminaries

1.1 HOW TO LIVE WITH COORDINATES

1.1.1 *Vectors, geometry*

When we are considering geographical data computers are both very good and very bad. They are very good at storing the connectivity between different objects – a graph in the mathematical sense. However, they are very bad at manipulating numbers being used as $X - Y$ coordinates. This is because the limited precision of computer arithmetic means that the calculation of intersections and similar entities are only ever approximate. This is fine if you only want to draw them, but if you wish to calculate whether a point is on one side of a line segment or the other, there is no guarantee that the result will be correct – or, even worse, whether it will be consistent when recalculated at a later time. For this reason computer scientists – computational geometers – have defined a small number of predicates that, if calculated correctly, may be used for a large number of individual geometrical operations. A great deal of work has been done on providing effective algorithms for these predicates, but the predicates that we will be using in two-dimensional Euclidean space may easily be calculated and have proved to be robust enough for most applications.

But first we must talk about coordinates. And the best way to talk about coordinates is to imagine that there aren't any: we are back in a simple world where we merely draw what we want to represent, like in the diagram below (Figure 1).

Here we have two arrows – vectors – merely drawn on the screen or paper. The arrows indicate that the head is at some position relative to the tail. But there are no Xs and Ys! Simple vector algebra states that vectors have length and orientation – but no position. This makes them free of any coordinate system being imagined – and even free of the number of dimensions.

But what can we do with them? They have length, so we can multiply them by some constant amount – called a scalar – while preserving the orientation (Figure 2). We can also multiply them by a negative scalar, changing the direction but keeping the overall 'orientation': this is equivalent to '**–a**' (Figure 3). (Of course, if we multiply them by zero all that remains is an oriented point!)

1.1.2 *Simple vector algebra*

We can also add and subtract vectors: to add vector **b** to vector **a**, just put the tail of **b** at the head of **a**. The result is the same as travelling along **a** and then along **b** – so we have a new vector **a** + **b** as shown in Figure 4. If we start with **b** and then add **a**, the result is **b** + **a** (the same as **a** + **b** because

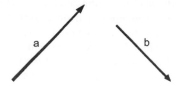

Figure 1.   Two vectors.

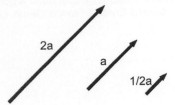

Figure 2. Scalar multiplication.

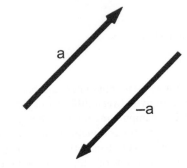

Figure 3. Orientation change.

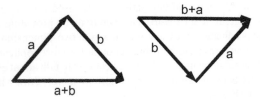

Figure 4. Vector addition.

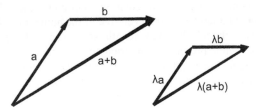

Figure 5. Add scaled vectors.

a vector's position doesn't matter). For subtraction, $\mathbf{a} - \mathbf{b} = \mathbf{a} + (-\mathbf{b})$, where $-\mathbf{b} = -1\mathbf{b}$. We can also multiply the whole operation by a scalar λ, (Figure 5).

If we want to relate this to the Cartesian coordinate system that we are familiar with, we put the tail of all vectors at the coordinate system origin, and the head at the XY location of the point. These are called position vectors. By convention a unit vector (a vector whose length is 1) along the x-axis is called '$\mathbf{i}$' and along the y-axis is called '$\mathbf{j}$'. (If we are working in three dimensions, the unit vector along the z-axis is called '$\mathbf{k}$'.) These are basis vectors. If we have a point P(x, y) with Cartesian coordinates (x, y) its position vector can be expressed as a linear combination $\mathbf{v} = (x, y) = x\mathbf{i} + y\mathbf{j}$. If we have a point P(x, y, z) with Cartesian coordinates (x, y, z) its position vector can be written as $\mathbf{v} = (x, y, z) = x\mathbf{i} + y\mathbf{j} + z\mathbf{k}$. Thus we have a link back to Cartesian coordinates when we need it.

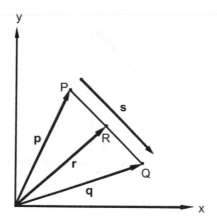

Figure 6. Mid-point.

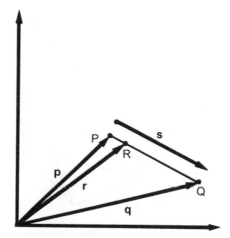

Figure 7. Parametric equation of a line.

If we have two position vectors **p** and **q** we could define a vector **s** (a directed line segment) from **p** to **q** as shown. Since **p** + **s** = **q**, clearly **s** = **q** − **p** (Figure 6).

Now this is where it starts getting useful. If **p** + **s** = **q** and we want half of the line segment **s**, then the new point R has position vector **r** = **p** + 1/2(**s**) = **p** + 1/2(**q** − **p**) = 1/2(**p**) + 1/2(**q**).

Now let **r** be anywhere along the line from P to Q – say $\lambda = 10\%$ of the way from P to Q. So **r** = **p** + λ**s** = **p** + λ(**q** − **p**), which equals $(1 - \lambda)$**p** + λ**q**. Take 10% of **q** and add 90% of **p** and you get **r** (Figure 7). If we let $\mu = 1 - \lambda$ the equation becomes **r** = μ**p** + λ**q**.

This gives us the parametric equation of a line. By an acceptable mixture of points and vectors, we obtain R = P + λ**s**. Given a single parameter λ, we can define any point along the line segment between P and Q. In fact, we can define any point along the infinite line through P and Q by letting λ take on any real value. And we can do it in any number of dimensions, not just two. Aren't vectors wonderful?

λ and μ are the barycentric coordinates for a 1D Simplex – a line: we are using 2 variables in a 1D space for consistency, and then adding that they sum to 1. We will see barycentric coordinates again later.

Thus $R_x = \lambda Q_x + (1 - \lambda)P_x$, and the same for R_y. If P and Q have some other attribute, such as colour or elevation (Z) then linear interpolation is performed by making $R_z = \lambda Q_z + (1 - \lambda)P_z$.

In Figure 8, if we construct vector **a** − **b** then from the cosine rule we have $\|\mathbf{a} - \mathbf{b}\|^2 = \|\mathbf{a}\|^2 + \|\mathbf{b}\|^2 - 2\|\mathbf{a}\|\|\mathbf{b}\|\cos\theta$.

3

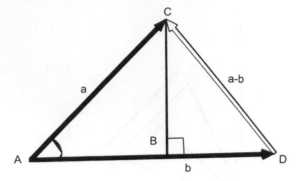

Figure 8.   Projection.

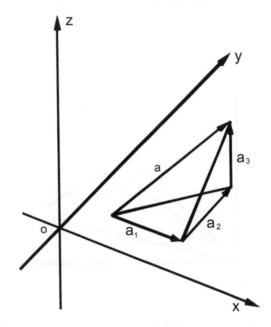

Figure 9.   Three dimensions.

Working with the coordinates, we can obtain: $\cos(\theta) = [a_1 b_1 + a_2 b_2]/[\|\mathbf{a}\|\,\|\mathbf{b}\|]$.

We define the dot product of **a** and **b** to be $\mathbf{a}.\mathbf{b} = a_1 b_1 + a_2 b_2$.

(In 3D, $\mathbf{a}.\mathbf{b} = a_1 b_1 + a_2 b_2 + a_3 b_3$, where a_1, a_2, a_3 are the first, second and third dimensions of **a** – and similarly for **b**.)

So $\cos(\theta) = \mathbf{a}.\mathbf{b}/\|\mathbf{a}\|\,\|\mathbf{b}\|$. This holds for any number of dimensions.

If $\cos(\theta)$ is positive then the distance AB is positive, and if $\mathbf{a}.\mathbf{b}$ is zero, then **a** is perpendicular to **b**.

By ordinary trigonometry $\cos(\theta) = \text{length (AB)}/\text{length (AC)}$, so the length of AB (the projection of **a** onto **b**) $= \|\mathbf{a}\|\cos(\theta) = \mathbf{a}.\mathbf{b}/\|\mathbf{b}\|$.

3D vectors

Vector operations are valid in any number of dimensions. In Figure 9, the vector **a** is composed of $\mathbf{a}_1 + \mathbf{a}_2 + \mathbf{a}_3$ along each of the coordinate axes.

Because vectors have no fixed position, if we want to work with vectors at arbitrary locations it is often convenient to move the starting point of the position vectors to the origin, as shown in Figure 10. This simplifies the calculations of areas or volumes, as described later.

4

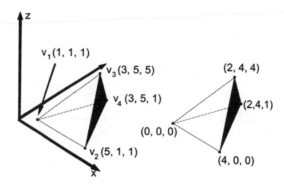

Figure 10.   Displacement to origin.

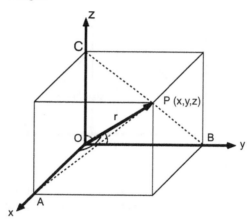

Figure 11.   Direction cosines.

For some applications 'direction cosines' are useful. The angles are formed in the plane containing the unit vector **r** of interest and the X, Y and Z axes. The cosines of these angles are the projections of the vector **r** onto the axes – and these are merely the x, y and z coordinates of the endpoint of **r** (Figure 11).

In 3D, two vectors **b** and **c** will always fall on a plane, and we often want to find a vector perpendicular to that plane – its 'normal' vector **n**. **n** will thus be perpendicular to **b** and to **c**. Thus $\mathbf{n.b} = 0$ and $\mathbf{n. c} = 0$. For ordinary simultaneous equations we would need a third equation to find an exact answer, but with only two we can only find a linear combination that satisfies our requirements – which is exactly what we need here: a normal vector, containing the proportions of the basis vectors **i**, **j** and **k**. This is most simply expressed as a determinant:

$$\mathbf{n} = \begin{vmatrix} \mathbf{i} & \mathbf{j} & \mathbf{k} \\ a_1 & a_2 & a_3 \\ b_1 & b_2 & b_3 \end{vmatrix} = \mathbf{a} \times \mathbf{b}$$

Vector **n** is perpendicular to vectors **a** and **b**.

Note that if θ is positive, going from **a** to **b** (anticlockwise), then **n** is oriented upwards (out of the paper) as on the left (Figure 12). If θ is negative (clockwise) this is equivalent to switching the second and third rows of the determinant (which, from the laws of determinants, changes the sign of the result) and **n** is oriented into the paper.

By expanding the above determinant we can calculate that the magnitude of the vector product $\mathbf{n} = \|\mathbf{a}\| \|\mathbf{b}\| \sin(\theta) = 2$.triangle-area (Figure 13).

5

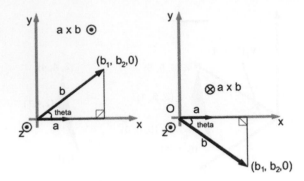

Figure 12. Orientation of cross product.

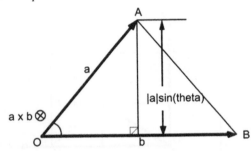

Figure 13. Magnitude of vector product.

If we know the vector normal **n** we know that any vector in the plane is perpendicular to **n**. Let **r** be any point in the plane, and let $\mathbf{r_0}$ be a fixed point we know. Then $(\mathbf{r} - \mathbf{r_0})$ is perpendicular to **n**, and $(\mathbf{r} - \mathbf{r_0}).\mathbf{n} = 0$.

We can rephrase this as $\mathbf{r.u} = \mathbf{r_0.u}$, where $\mathbf{u} = \mathbf{n}/\|\mathbf{n}\|$. Since we know **u** and we know $\mathbf{r_0}$, we can calculate an equation in (x, y) for **r**, giving us an equation for the plane.

1.1.3 *CCW predicates (2D)*

The two most common 'multiplication' operations are the scalar product and the vector product: combining these produces the triple scalar product, or box product, $\mathbf{a} . \mathbf{b} \times \mathbf{c} = \mathbf{a} . \mathbf{d}$ (Figure 14).

In the first step we calculate the vector normal **d** of vectors **b** and **c** ($\mathbf{d} = \mathbf{b} \times \mathbf{c}$). Its magnitude equals the area of the parallelogram COB.

Then we project vector **a** onto **d**: $h = \|\mathbf{a}\|\cos\theta$ (angle between **a** and **d**), where $\cos\theta = \mathbf{a.d}/\|\mathbf{a}\|\|\mathbf{d}\|$ from the scalar product described above.

So $h = \mathbf{a.d}/\|\mathbf{d}\|$ and the volume $= h\|\mathbf{d}\|$

but $\|\mathbf{d}\| = \mathbf{b} \times \mathbf{c}$

so the volume is $\mathbf{a.b} \times \mathbf{c}$.

This is a scalar value and is 6*volume given by $\mathbf{a.b} \times \mathbf{c}$.

For three vertices (x_1, y_1, z_1), (x_2, y_2, z_2) and (x_3, y_3, z_3) this may be conveniently written as a determinant:

$$D = \begin{vmatrix} x_1 & y_1 & z_1 \\ x_2 & y_2 & z_2 \\ x_3 & y_3 & z_3 \end{vmatrix} = x_1.\begin{vmatrix} y_2 & z_2 \\ y_3 & z_3 \end{vmatrix} - y_1.\begin{vmatrix} x_2 & z_2 \\ x_3 & z_3 \end{vmatrix} + z_1.\begin{vmatrix} x_2 & y_2 \\ x_3 & y_3 \end{vmatrix}$$

$$= x_1.(y_2.z_3 - y_3.z_2) - y_1.(\ldots) + z_1.(\ldots)$$

Now let us rotate the points A, B and C so they lie on a horizontal plane, and we will set the z values of all the points to 1. (Remember, rotation doesn't change the vector results.)

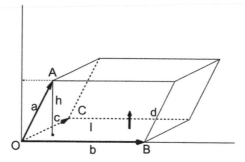

Figure 14.   The box product.

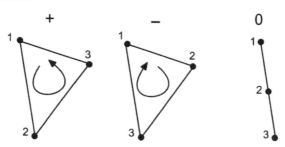

Figure 15.   CCW predicate.

Thus the volume = (base area in x, y)*1 = (area of the triangle ABC)*2.

The result is a scalar – a single numerical value – that takes the input x, y coordinates of three vertices and returns the double-area of the triangle they define. (In 3D, with four vertices and three coordinate axes, it returns 1/6 of the volume of the tetrahedron formed.) It is calculated as the determinant:

$$CCW = \begin{vmatrix} x_1 & y_1 & 1 \\ x_2 & y_2 & 1 \\ x_3 & y_3 & 1 \end{vmatrix}$$

As with all determinants, if we exchange any two rows or columns the sign of CCW changes: if V_1, V_2, V_3 form an anticlockwise triangle then CCW is positive; if we exchange two rows then the triangle is formed as V_1, V_3, V_2, and CCW has a negative sign. This predicate is called 'CCW' – for Counter-Clock-Wise – and is used to test the orientation of the triangle. If the three vertices fall on a straight line then CCW = 0 (Figure 15).

This 'CCW' is a 'signed area' or volume, and is a robust predicate of computational geometry. If we exchange two rows, or two columns, the sign of CCW changes. This can tell us if we have an anticlockwise triangle (CCW +ve), or clockwise (CCW –ve). Much work has been spent in making it robust – that is, giving consistent answers in all circumstances. However, for many applications the calculation of the simple determinant suffices.

We previously looked at the parametric equation of a line. We can use CCW to define the parametric equation of a triangle, simplifying planar interpolation, walking through a triangulation, transforming map coordinates, simplifying line intersection tests and other useful stuff.

For us the most valuable result of basic vector algebra is the triple scalar (box) product – it gives us the signed area of a triangle, which is the basic predicate for much computational geometry. (My students used to call this determinant the 'Magic Matrix' – no matter the question, the answer was CCW!)

With care, the single-precision determinant suffices to test simple triangles – the important thing is to ensure you always get the same result for the same triangle – always go anticlockwise and start at the same vertex.

7

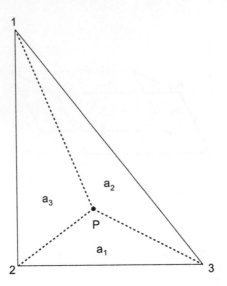

Figure 16.　Areal coordinates.

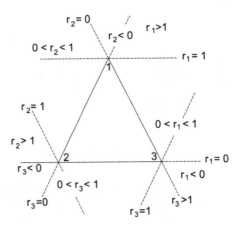

Figure 17.　Barycentric coordinates.

1.1.4　*Sidedness, barycentric coordinates*

CCW thus expresses the relationship of a point P with the end points of a line V_1 V_2: if P corresponds to V_3 of a triangle then the CCW magnitude gives the double-area of the triangle, and the sign gives the triangle orientation. Consider, however, the case where P is interior to the triangle: if CCW (V_1, V_2, V_3) is positive then CCW (V_1, V_2, P) will also be positive (Figure 16).

The same will hold true for the other two edges – this test determines if P is inside the triangle and, if not, a negative sign indicates the edge it falls outside. These three values form a coordinate system (a_1, a_2, a_3) based on the triangle. If they are normalized by dividing by CCW (V_1, V_2, V_3) they are referred to as barycentric or 'areal' coordinates (r_1, r_2, r_3) and sum to 1: we have converted from global coordinates (e.g. of the type used in raster models) to a system based on objects (our three vertices) (Figure 17).

8

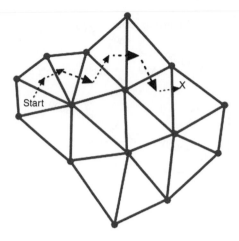

Figure 18. Triangle walk.

Several useful applications follow from this. The first is that (r_1, r_2, r_3) give a linear combination of the three vertices: multiply (x_1, x_2, x_3) by their corresponding r values and you obtain x_P, the X coordinate of P used in the determinant CCW:

$$CCW = \begin{vmatrix} x_1 & y_1 & 1 \\ x_2 & y_2 & 1 \\ x_3 & y_3 & 1 \end{vmatrix}$$

$$r_1 = \begin{vmatrix} x_P & y_P & 1 \\ x_2 & y_2 & 1 \\ x_3 & y_3 & 1 \end{vmatrix} /CCW \text{ and similarly for } r_2 \text{ and } r_3.$$

Then:

$$x_P = r_1 x_1 + r_2 x_2 + r_3 x_3$$

$$y_P = r_1 y_1 + r_2 y_2 + r_3 y_3$$

which returns us to our original coordinate system, and

$$z_P = r_1 z_1 + r_2 z_2 + r_3 z_3$$

where z_P may be some attribute of the data at location (x_P, y_P) – perhaps elevation. Thus we can interpolate linearly within our triangle, giving us a planar triangular surface as part of our terrain model.

(Incidentally, this also works in 3D: four CCW calculations, using X, Y and Z for the four vertices of a tetrahedron, positions our point P with respect to the tetrahedron and gives us volumetric interpolation.)

This is very useful if we are working with a TIN – a terrain model composed of triangles – but we have the additional job of determining the triangle containing P. Assuming we have a data structure that tells us which triangles are adjacent to which, we can use CCW first of all to find if P is 'inside' all three edges of some starting triangle and, if it is not, to determine which edge(s) it is outside: then we go to the appropriate adjacent triangle and repeat the exercise until (r_1, r_2, r_3) are all positive – that is the enclosing triangle. Having got there, use (r_1, r_2, r_3) to estimate z_P (at the location of P) (Figure 18): job finished. This is referred to as a triangle 'walk'.

This of course assumes that the set of triangles is known: there are many ways to triangulate a set of points: which way is best? Various approaches have been tried, but all but one have a serious

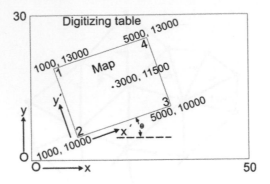

Figure 19.   Map digitizing.

flaw: there is no guarantee that the same triangulation will result if the points are processed in a different order – many of the triangles may differ, for example, if just one point is left out or displaced. This may be phrased differently: adding, deleting or moving a single point should have only a local effect. The exception to this, and hence the one that is used in most TIN software, is the Delaunay triangulation (DT): this may be derived conceptually from the Voronoi diagram (VD): these will be discussed later.

Another example where barycentric coordinates based on the triangle may be useful is in coordinate transformations of a map – perhaps when it is being digitized. Typically it is taped onto a digitizing table, approximately aligned with the axes, and control points are added manually – clicking on the map control point to get the 'table' coordinates and then typing in the real 'map' coordinates – perhaps based on a map projection such as UTM. Traditionally, 2D computer graphics transformations (described briefly below) are then applied to align the map axes with the table, to superimpose the origins of the two coordinate systems and to scale them. This can have various problems, such as when the map axes are not precisely perpendicular (often due to paper shrinkage). Barycentric coordinates provide a convenient alternative.

Perhaps the corners of the map provide the required control points: here points P_1, P_2 and P_3 give our base triangle (Figure 19). Any point P digitized in table coordinates may have its areal coordinates r_1, r_2 and r_3 determined as above, giving the position of point P relative to these vertices – that is, in the 'source' (table) coordinate system. These proportions will also be true in the 'target' (map) system – so we simply convert back from r_1, r_2, r_3 to x_P and y_P using the map coordinates of our control points:

$$x_P(\text{map}) = r_1 x_1(\text{map}) + r_2 x_2(\text{map}) + r_3 x_3(\text{map})$$

$$y_P(\text{map}) = r_1 y_1(\text{map}) + r_2 y_2(\text{map}) + r_3 y_3(\text{map})$$

In this way we need not be concerned about axis alignment, origins or scale changes, and all transformed points are a linear interpolation within (or outside) the base triangle. Clicking on a control point will give a correct result, and points along the triangle edges will be transformed linearly and would match map transformations based on adjacent control triangles. Note however that this does not apply to point P_4, as this was not used as a control point: if we are doing series mapping, with many adjacent map sheets, additional adjustments must be applied so that transformation of point P_4 matches its map coordinates precisely. (We do this by calculating r_1, r_2 and r_3 for point P_4, just as we do for any other point, in both table and map coordinate systems, and calculating the ratios in each case. These are then used to 'adjust' our original three-point areal coordinates, so that they will now give the right answer at point P_4. See Wolberg (1994).) Consequently, for the three point correction alone, the mid-point of the map in the figure below will not necessarily fall precisely at the intersection of the map diagonals.

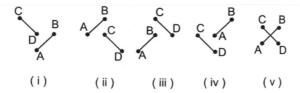

(i) (ii) (iii) (iv) (v)

Figure 20.   Intersection tests.

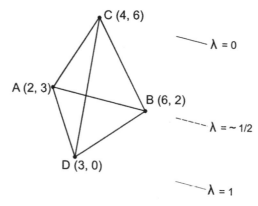

Figure 21.   Intersection calculation.

In many cases several points interior to the map may be used as control points if their precise map coordinates are known. These may then be triangulated, and for each digitized point P the appropriate base triangle may be found, as above, and the map transformation applied. The result is a piecewise linear transformation so, for example, a straight road digitized in one base triangle will connect with its continuation in the next, but there may well be a bend in it.

Finally, 2D barycentric coordinates are particularly useful in the fundamental cartographic question of the detection and calculation of line segment intersection. Traditionally this is solved by finding the linear equation of the type 'y = mx + c' for each segment defined by two end-points, and then solving the simultaneous equations to find y at the intersection point. This has several problems: the solution is unstable or does not exist for two nearly-parallel lines; the slopes 'm' may be nearly infinite if the line is near vertical; and the intersection may fall outside one or both line segments. Barycentric coordinates significantly simplify the problem.

For two line segments AB and CD, calculate CCW for ABC, ABD, CDA and CDB – the first two points forming one line segment and the third being taken from the other. These determinants may be positive or negative. If (say) ABC and ABD have the same sign then CD falls entirely on one side or the other of AB. Thus no intersection is possible. If CDA and CDB have the same sign then AB is entirely on one side or the other of CD. In either of these cases no intersection is possible and the test terminates.

If any of the tests return a 0 then one end of a segment falls on the (maybe extended) line of the other; if both are zero we have a zero-length line segment and no intersection is possible. In Figure 20, case (i) has CD to the left of AB (ABC and ABD are both positive), and case (ii) has it on the right (both ABC and ABD are negative). In case (iii) AB is to the right of CD (both CDA and CDB are negative), and in (iv) AB is to the left (both CDA and CDB are positive). Case (v) shows where each pair of CCWs are of opposite sign and an intersection may be calculated.

The intersection point may be calculated from these same determinants, as in Figure 21. Let λ represent the parametric value of the intersection of AB along CD. Triangles ABC and ABD have areas proportional to their heights, as they both have the same base AB – and CCW gives these

11

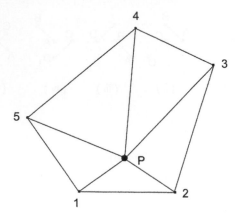

Figure 22.   Convex polygon area.

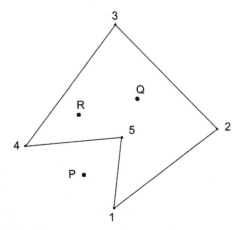

Figure 23.   Point in concave polygon.

areas (x2). Let $\lambda = $ ABC/(ABC-ABD) – this will equal 0 when the intersection is at point C, and 1 when at point D. (One of these CCWs will always be positive and the other negative, although either may be 0. If both are 0 then both segments are in-line and no intersection calculation is possible.)

Then: $x_{int} = \lambda x_D + (1 - \lambda)x_C$ and similarly for y – however, for some problems only the existence of an intersection needs to be known. This is much faster, and simpler, than the traditional approach, and gives robust results.

1.1.5   *Area, volume, intersection calculations*

A very simple and useful application of the 'signed area' is in the calculation of polygon areas. This presumes that the vertices of the polygon are labelled in order – and, based on the CCW predicate, in anti-clockwise order. Starting at some origin point P, triangle areas are calculated as P, V_i, V_{i+1} and the resulting areas summed. This will give the (double) area of the final polygon.

When trying to find if a point falls within a given polygon, it does so for a convex one if all the triangle areas are positive (Figure 22). For a concave polygon this may not give the correct answer: in Figure 23, point Q is on the 'positive' side of all the edges (and is indeed inside) but R is on the wrong side of side 5 – 1 (although it is inside). Point P (outside) is on the negative sides of edges 4 – 5 and 5 – 1. More elaborate tests are needed in these cases, involving either the Jordan Curve Theorem or the Line Segment Voronoi Diagram (to be discussed later).

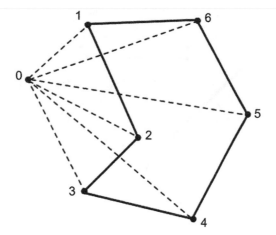

Figure 24.   Polygon area.

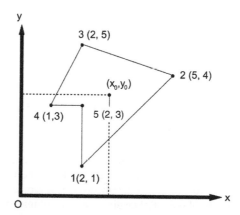

Figure 25.   Polygon centroid.

It may well be desirable to use the origin of the coordinate system to calculate the areas, as then the determinant is reduced to 2×2:

$$CCW = \begin{vmatrix} 0 & 0 & 1 \\ x_2 & y_2 & 1 \\ x_3 & y_3 & 1 \end{vmatrix} = \begin{vmatrix} x_2 & y_2 \\ x_3 & y_3 \end{vmatrix} = x_2\,y_3 - x_3\,y_2$$

In Figure 24, 0, 1, 2 and 0, 2, 3 are clockwise triangles, and therefore have negative areas, while 0, 3, 4 through 0, 6, 1 are anticlockwise and therefore positive. The total sum of the positive parts and the negative parts gives the polygon area.

The 'centroid' or centre of gravity (x_0, y_0) of the polygon may be found by taking the centroid of each triangle (the average of its vertices), weighting the coordinates by the triangle area, and summing the result for X, and then for Y (Figure 25). This works for concave as well as convex polygons, but the resulting centroid may not always fall within the polygon – try balancing a doughnut (with a hole) on a pencil!

If areas are to be calculated for polygons with islands the simplest way is to 'cut' a path from the inside to the outside and then follow the resulting modified boundary: the hole will be traversed in a clockwise direction (Figure 26).

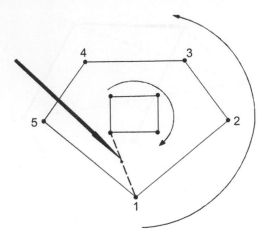

Figure 26.   Polygon with a hole.

1.1.6   *3D sidedness and intersection*

CCW can be extended directly to (x, y, z, 1):

$$3D\ CCW = \begin{vmatrix} x_0 & y_0 & z_0 & 1 \\ x_1 & y_1 & z_1 & 1 \\ x_2 & y_2 & z_2 & 1 \\ x_3 & y_3 & z_3 & 1 \end{vmatrix} = 6 \times \text{the volume of a tetrahedron.}$$

This gives the 'signed' volume of the tetrahedron (4 vertices) rather than the signed area of a triangle in 2D. These 'volume coordinates' may be used in the same way as 2D areal coordinates to locate a point P with respect to a reference tetrahedron: just replace each vertex in turn in the determinant with the coordinates of point P, then divide by the volume of the reference tetrahedron. The result is a set of four barycentric coordinates (adding up to 1) that vary from 0 if P is on the plane of the tetrahedron face opposite the vertex that has been replaced, up to 1 at the parallel plane through the vertex itself. If P is below the base face then 3D CCW is negative.

This locates P with respect to the four vertices, easily determining if P is inside the tetrahedron. As in 2D, linear interpolation in 3D may be performed by taking the weighted average of the attributes at the vertices times the appropriate barycentric coordinates. This becomes convenient when interpolating 3D data such as sea temperature or salinity.

The volume of a polyhedron may be calculated, as for 2D polygons, by taking a reference point and summing the volumes of the tetrahedra formed by each face in turn plus the reference point. Care must be taken that all triangular polyhedron faces are defined in the same order – usually anticlockwise when looking at the face from the polyhedron interior. Centroids may be calculated in the same way. 3D coordinate transformations, as in 2D map transformations, may be performed when we have coordinates in both the source and target systems at each control point.

Following the rules for the direction of the vector normal, tetrahedron vertices must be ordered appropriately. In Figure 27 vertices V_0, V_1 and V_2 form an anticlockwise triangle, and the normal vector to vectors (0, 1) and (0, 2) projects upwards. Thus the vector normal, and the 3D CCW, are positive upwards.

The 3D CCW allows the relatively simple determination of a variety of 3D geometric tests that would be awkward otherwise. A good example is the question of the intersection of a line segment with a triangle in 3D. In Figure 28 vertices V_P, V_Q and V_R form a triangle (anticlockwise upwards) with the vector normal pointing upwards. 3D CCW (V_P, V_Q, V_R, V_A) thus has a positive value/volume, and 3D CCW (V_P, V_Q, V_R, V_B) has a negative one. As with the 2D line segment

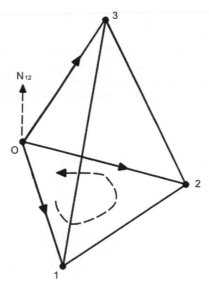

Figure 27.   Tetrahedron vertex ordering.

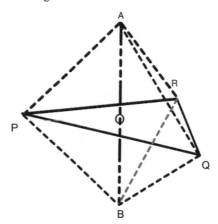

Figure 28.   3D intersection of line segment and triangle.

intersection test, this means that AB crosses the plane of PQR – if they both had the same sign then no intersection would exist.

The next question is to determine if the intersection point falls within the triangle itself. Again looking at the figure, we can examine segments AB and PQ: calculating 3D CCW (V_P, V_Q, V_A, V_B) will give a signed volume for that tetrahedron. We do not necessarily know if it is positive or negative – and in any case the sign would change if we switched the endpoints A and B. However, we do know that the sign would change if segment AB moved to the other side of PQ – and it would be precisely 0 if the two segments intersected exactly in 3D space.

We also know that if we move round the triangle PQR and calculate 3D CCW (V_Q, V_R, V_A, V_B) instead of 3D CCW (V_P, V_Q, V_A, V_B) the same conditions would hold: the sign for 'inside' would be the same as before – and the same would be true for 3D CCW (V_R, V_P, V_A, V_B). Thus if all three signs were the same, the intersection point would be inside triangle PQR.

Finally, if the precise intersection coordinates are required, λ (the parameter along AB) would be calculated as for 2D, but using the ratios of 3D CCW (V_P, V_Q, V_R, V_A) and 3D CCW (V_P, V_Q, V_R, V_B). The coordinates x, y, z would be calculated from λ as before, using the coordinates of A and B.

This should give a taste of the uses of 3D CCW – many others can be imagined. Barycentric coordinates are very powerful tools.

We started with barycentric coordinates in 1D – the parametric equation of a line. In 2D, areal coordinates were used for many questions concerning triangles and the relative positions of vertices, as well as line segment intersection. In 3D the relative positions of points, line segments and triangles help solve otherwise awkward problems. In all three cases linear interpolation is readily addressed.

FURTHER READING

Most people learn basic vector algebra at school, but it often doesn't stick – maybe because it is treated algebraically without much in the way of geometric figures, or maybe because one stops with the simple theorems without continuing to see how useful they can be in the 'real world'. The mathematician Roger Penrose once said that mathematicians either think in algebra or else in geometry – and he initially thought in geometry. Taking encouragement from that, we look at pictures as far as possible, and then move as fast as we can to useful applications – especially the ones we need in order to look at spatial relations.

For reading up on any aspect touched on rather lightly here, any introductory textbook on vector algebra will do, but as this book emphasizes the geometrical (graphical), rather than the algebraic, approach it would be good to pick one with the same emphasis. My battered old copy of Faux and Pratt (1979) 'Computational Geometry for Design and Manufacture' has served me well. (Beware! 'Computational Geometry' has taken on a new meaning since then.) Whichever book you choose, don't read the whole 'Table of Contents' and get scared by advanced tensors – we really only need the first couple of chapters in most cases. Another book I have read, and re-read, from cover to cover, is Sedgewick (1983) 'Algorithms', where an 'algorithm' is a recipe or procedure, typically to produce a particular data structure.

Most of us have problems handling geometric tests in 3D – because we can't draw them! However, vector operations are independent of coordinate systems – and hence of the number of dimensions. Thus CCW extends directly, and permits a variety of 3D queries and intersection tests. For nasty cases Shewchuk's CCW Predicate can be used – see Shewchuk (1997).

Early discussions on 2D intersection tests include Douglas (1974), and Saalfeld (1987). Huang and Shi (1997) reviewed the point-in-polygon problem. Wolberg (1994) describes image (or map) warping.

1.1.7 *Computer graphics*

We have considered how to use coordinates to determine the relationships between points, lines and areas (triangles) to manipulate our model of real-world data. A final consideration that requires coordinates is: how to display the results.

While spatial relationships may best be manipulated using local, relative coordinates, the display of the whole model inevitably requires a global coordinate system – almost invariably Cartesian coordinates. These usually follow the form we encountered in our vector algebra: the Y axis perpendicular to the X axis, and with an anticlockwise orientation. As we have seen, the vector normal of these two base vectors is perpendicular to them both, and pointing 'out' of the page – (Figure 29). Angles are positive in an anticlockwise sense, between X and Y, Y and Z, and Z and X. This is a 'right-handed' system.

Warning: when we are considering the X-Y coordinates on a screen it appears natural to define the positive Z axis as pointing into the screen: this is a left-handed coordinate system! (Figure 30).

If we wish to manipulate our model with respect to the viewpoint of the observer (usually at a short distance in front of the screen) we need primarily to be able to move ('translate') them, rotate them, and scale them. This can always be done on a case-by-case basis with standard coordinate geometry and trigonometry, but a much improved approach, and the one now universally used in computer graphics, is to frame these operations in terms of matrix algebra operations, so that

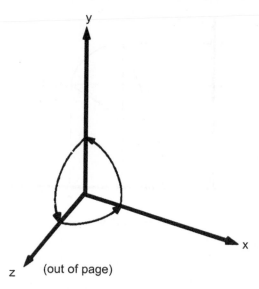

Figure 29.   Right handed coordinate system.

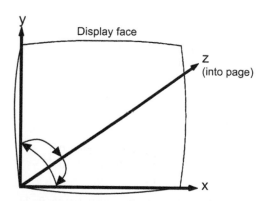

Figure 30.   Left handed coordinate system.

multiple operations (move, scale, rotate, move again ...) may be combined as a single matrix applied to the coordinates of the model – matrix multiplication operations make this straightforward.

We can express the rotation of a point around the origin by providing a rotation matrix:

$$\begin{bmatrix} x' & y' \end{bmatrix} = \begin{bmatrix} x & y \end{bmatrix} \begin{bmatrix} \cos\theta & \sin\theta \\ -\sin\theta & \cos\theta \end{bmatrix} \quad \text{Rotate CCW by } \theta$$

and scaling it (again around the origin):

$$\begin{bmatrix} x' & y' \end{bmatrix} = \begin{bmatrix} x & y \end{bmatrix} \begin{bmatrix} S & 0 \\ 0 & S \end{bmatrix} \quad \text{Scale by } S$$

Note: if we want to rotate or scale a point on an object around perhaps some central point we need first of all to move that object's centre to the origin, apply the transformation matrix, and then move the centre back.

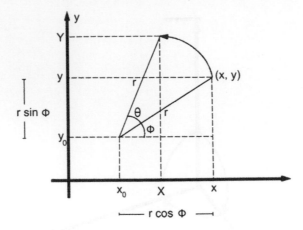

Figure 31. Point rotation.

But how do we move a point (add something to x and y)? – We cannot do this with a 2×2 matrix. However, if we add a dimension (usually called 'w') and put our 2D model on the $w = 1$ plane – we can do it:

$$[x' \quad y' \quad 1] = [x \quad y \quad 1] \begin{bmatrix} 1 & 0 & 0 \\ 0 & 1 & 0 \\ dx & dy & 1 \end{bmatrix} = [x + dx, \; y + dy, \; 1] \quad \text{Translation}$$

The other transformations work fine too:
  Rotate (see Figure 31)

$$[x' \quad y' \quad 1] = [x \quad y \quad 1] \begin{bmatrix} \cos\theta & \sin\theta & 0 \\ -\sin\theta & \cos\theta & 0 \\ 0 & 0 & 1 \end{bmatrix}$$

$$= [x\cos\theta - y\sin\theta, \; x\sin\theta + y\cos\theta, \; 1]$$

$$[x' \quad y' \quad 1] = [x \quad y \quad 1] \begin{bmatrix} 0 & S & 0 \\ 0 & S & 0 \\ 0 & 0 & 1 \end{bmatrix} = [Sx \; Sy \; 1] \qquad \text{Scale}$$

Here is an example of a 'house' (Figure 32) being rotated $+ 90$ degrees around its centre $(2, 2)$ and being scaled by 2 (Figure 33).
  The same thing can be done in 3D (Figure 34):

$$[x' \quad y' \quad z' \quad 1] = [x \quad y \quad z \quad 1] \begin{bmatrix} S & 0 & 0 & 0 \\ 0 & S & 0 & 0 \\ 0 & 0 & S & 0 \\ 0 & 0 & 0 & 1 \end{bmatrix} \qquad \text{Scale}$$

$$[x' \quad y' \quad z' \quad 1] = [x \quad y \quad z \quad 1] \begin{bmatrix} 1 & 0 & 0 & 0 \\ 0 & 1 & 0 & 0 \\ 0 & 0 & 1 & 0 \\ dx & dy & dz & 1 \end{bmatrix} \qquad \text{Translate}$$

18

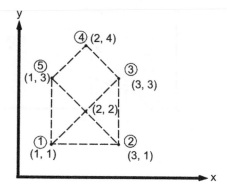

Figure 32. House.

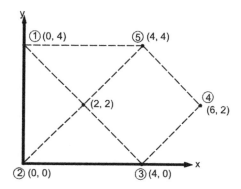

Figure 33. Rotated house

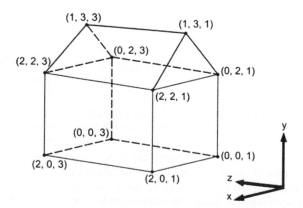

Figure 34. 3D house.

$$[x' \quad y' \quad z' \quad 1] = [x \quad y \quad z \quad 1] \begin{bmatrix} 1 & 0 & 0 & 0 \\ 0 & 1 & 0 & 0 \\ 0 & 0 & 1 & 0 \\ 0 & 0 & 0 & 1 \end{bmatrix} \quad \text{Identity (no change)}$$

but there are three different rotations, one about each axis.

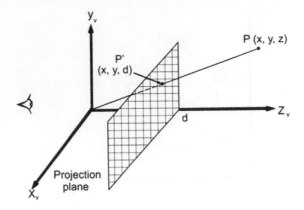

Figure 35. Perspective.

3D Transformation with a 4 × 4 matrix

$$[x' \quad y' \quad z' \quad 1] = [x \quad y \quad z \quad 1] \begin{bmatrix} \cos\theta & \sin\theta & 0 & 0 \\ -\sin\theta & \cos\theta & 0 & 0 \\ 0 & 0 & 1 & 0 \\ 0 & 0 & 0 & 1 \end{bmatrix} \quad \text{Rot(Z)}$$

$$[x' \quad y' \quad z' \quad 1] = [x \quad y \quad z \quad 1] \begin{bmatrix} 1 & 0 & 0 & 0 \\ 0 & \cos\theta & \sin\theta & 0 \\ 0 & -\sin\theta & \cos\theta & 0 \\ 0 & 0 & 0 & 1 \end{bmatrix} \quad \text{Rot(X)}$$

$$[x' \quad y' \quad z' \quad 1] = [x \quad y \quad z \quad 1] \begin{bmatrix} \cos\theta & 0 & -\sin\theta & 0 \\ 0 & 1 & 0 & 0 \\ \sin\theta & 0 & \cos\theta & 0 \\ 0 & 0 & 0 & 1 \end{bmatrix} \quad \text{Rot(Y)}$$

Thus if we add our extra dimension (the '1', usually called 'w') to the matrices then we can produce a matrix that adds 'dx', 'dy' (and 'dz') to our object. These are 'homogeneous coordinates' – similar to the idea we used before for CCW. In fact, there is a direct transformation between the two, as in the map digitizing case above. We 'forward' calculate r_1, r_2 and r_3 in the 'table' coordinates x, y, z (where z always equals 1, as for our derivation of CCW). By implication, our control points are $V_1 = (1, 0, 1)$, $V_2 = (0, 1, 1)$ and $V_3 = (0, 0, 1)$. To 'back' calculate, we use the areal coordinates to determine P_x etc. (Barycentric (or 'areal') coordinates are homogeneous, with the added constraint that they sum to 1 – see Coxeter, 1969).

We can now combine (concatenate) our transformation matrices before applying them, allowing us to keep just one copy of our original object, plus one transformation matrix for each copy (instance) being displayed, no matter how many individual movements were involved.

Finally, 3D graphics requires a perspective view.

We have 4 × 4 matrices for scaling, translation and rotation around the X, Y and Z axes.

Perspective transformations assume the observer is at a distance 'd' in the system (see Figure 35, Figure 36, Figure 37), so x' = x(d/z), y' = y(d/z) and z' = z (d/z) = d.

(We can put this in the stack of matrices by taking the identity matrix and putting '1/d' in the 'w' position of the 'Z' row. When multiplied with the other matrices this means all 'w's = 'z/d' and

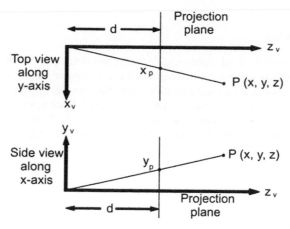

Figure 36.   Perspective from above.

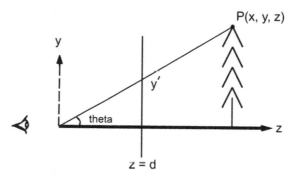

Figure 37.   Perspective from the side.

the other values are unchanged. However, because w is no longer 1, before we draw anything we must divide by this value (called 'perspective division') before drawing on the screen.)

A 3D graphics card (or a software simulation) therefore has (at least) the following stages:

a) Calculate the new rotation, scaling and translation matrices for each object instance.
b) Combine them to get a 'Transformation Matrix' T.
c) Combine this with the previous matrix T, if any.
d) Combine this with the perspective matrix.
e) Perform perspective division.
f) Apply this to each vertex of the object to project it onto the screen.

All objects have a transformation matrix showing translation, rotation, and scaling with respect to an 'origin'. This is a hierarchy, so you can put a wheel at the corner of a car, etc. for visible objects, and rotate it about the local axle, for example. Most graphics systems apply this principle to lights and cameras as well, so you can have a light from a lighthouse at a particular location, as well as a camera (view) on the bridge of a ship or elsewhere – and change it as required. This is a simple form of a 'Scene Graph'.

This book is intended to emphasize the geometric, rather than the algebraic, examination of spatial problems, and so as far as possible Cartesian coordinates are restricted to data input and data display. The above matrix algebra is included to illustrate the wide variety of applications where homogeneous coordinates are appropriate.

FURTHER READING

My first computer graphics book, read multiple times, was Foley et al. (1990), but a more recent book, Angel (2006), very conveniently integrates basic computer graphics operations with the OpenGL language available on most graphics cards. Coxeter (1969) first defined barycentric coordinates. Phong (1975) first outlined computer generated shading, and Blinn (1977) outlined the concatenated homogeneous coordinates described above – both of these are now integrated within OpenGL, as well as DirectX. Gold et al. (2004) describe a simple scene graph.

1.2   GRAPHS

1.2.1   *Graphs, overview: Planar graphs, duality*

Coordinates are difficult to handle in a computer – so we have tried to reduce access to them to a few simple predicates, together with output display. On the other hand, once we can talk about spatial relationships we may access tools that are well handled in computers – the manipulation of graphs.

A graph consists of a set of vertices and connecting edges: a connected graph, the usual case, has no disconnected portions. Graphs, like vectors, have no coordinates associated with the vertices. However, the edges of a graph may be given a direction, like one-way streets, and the edges (and sometimes the vertices) may be assigned specific weights. These are used in various forms of analysis.

While graphs in general have no fixed dimensionality, a form of graph that is particularly useful is the planar graph – a graph that, when drawn on a 2-manifold (such as a balloon) or a piece of paper may be represented without any of the edges crossing each other. (A non-planar graph is one that cannot be drawn without some edges crossing.) A planar graph therefore may have regions assigned to minimal closed loops of vertices and edges. The mathematician Kuratowski showed that any graph may be built from a planar graph and the two graphs K(3,3) (Figure 38) and K(5) (Figure 39). Attempting to redraw them so no edges cross is not possible, as illustrated for K(5).

Euler's formula for planar graphs states that $V + F - E = 2$, where V, F and E are the numbers of vertices, faces and edges respectively. The 'exterior' of the graph counts as a face – this is most easily seen if you draw your graph on a balloon! In Figure 40 this is: $5 + 5 - 8 = 2$.

A 2-manifold is a (2D) surface in 3D (or higher) space where every location can be thought of as a small 2D disc. Examples are a balloon or a doughnut, but not a box with many compartments.

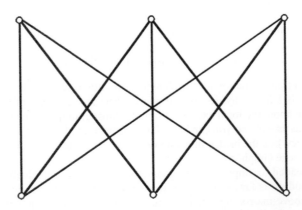

Figure 38.   K(3,3).

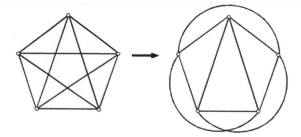

Figure 39.   K(5).

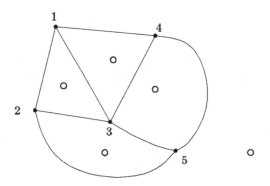

Figure 40.   A planar graph.

Its genus is the number of holes in the bounded object: 0 for a balloon, 1 for a doughnut (torus), etc. The Euler-Poincaré formula: $V + F - E = 2 - 2G$ allows for this.

The dual of a planar graph may most easily be thought of as follows:

If we start with a set of regions (adjacent countries perhaps), put a 'capital city' within each, and connect the capitals whenever two countries have a common boundary, the result is a dual graph.

(In most cases only three countries will meet at a junction, giving us a dual triangulation, but if four or more regions meet at the node then the dual may have four or more edges – which can be split up into triangles if required.)

This dual graph is also composed of vertices and edges: the vertices represent the original regions and the edges were defined to cross the original region boundaries. Both the primal and dual graphs follow Euler's formula, that $V + F - E = 2$, where $V =$ the number of vertices, $F =$ the number of regions (including the exterior), and $E =$ the number of edges (Figure 41).

Thus every region in one graph becomes a node in the other, every node becomes a region, and each edge becomes another edge that crosses the original one. Calculating the dual graph of a dual graph reproduces the original one. The dual graph of a tessellation of polygons is usually a triangulation: faces may have four or more edges if the original generating graph has a node with four or more adjacent nodes (Figure 41).

1.2.2   Graph traversal

Many query types for graphs involve examining each node – and we usually wish to traverse each edge exactly once: this is 'Eulerian graph traversal'. (Hamiltonian graph traversal, on the other hand, aims to visit each node precisely once: this is more difficult.) This is applied to a planar graph.

Traversal of a planar graph consists of converting a graph to a tree (a graph with only one region); this is achieved by starting at a 'root' node and visiting each connected node – but only if it has

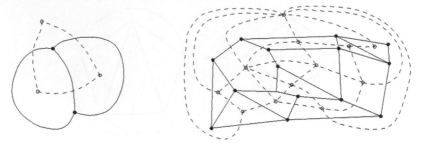

Figure 41.   Dual graphs.

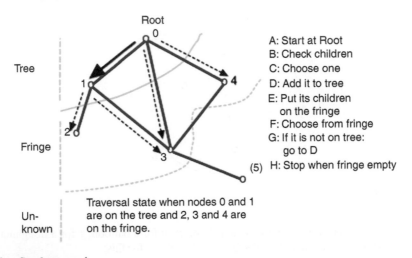

Root

Tree

Fringe

Un-
known

A: Start at Root
B: Check children
C: Choose one
D: Add it to tree
E: Put its children
   on the fringe
F: Choose from fringe
G: If it is not on tree:
   go to D
(5) H: Stop when fringe empty

Traversal state when nodes 0 and 1
are on the tree and 2, 3 and 4 are
on the fringe.

Figure 42.   Graph traversal.

not previously been visited. This means that, in the end, some edges will be ignored. A simple algorithm, appropriate for various types of traversals, is the Fringe Algorithm. It is applicable to simple undirected graphs, directed graphs where edges have directions, and weighted graphs where edges have a weight value (perhaps length, travel time or importance).

The graph is first stored in some appropriate manner. For a small one it may be stored in a matrix with rows representing the 'from' nodes and columns the 'to' nodes. Each cell represents a connection between a 'from' node and a 'to' node. This matrix will usually be sparse, with most cells empty. Other, more efficient, techniques are available for large graphs. A root node is selected, and placed on the 'tree' or final graph, and its connected neighbours – its 'to' nodes – are placed on the fringe and then examined in turn. One of them is selected on the basis of the type of transversal required, and it – along with the edge connecting it to its parent – is placed on the tree. Its own connected neighbours are then added to the fringe. One of the available nodes-plus-parent-edges is then selected, and put on the tree, and the process is repeated until no more nodes are available: the fringe is empty and the whole graph has been processed.

This algorithm is only appropriate for a connected tree, where the whole graph is connected by edges. One additional step is required: when an edge is selected from the fringe it must be tested to see if it connects to a new node and not one already on the tree, otherwise closed loops will be formed and the desired tree structure destroyed (Figure 42).

The remaining question is: how to choose the available edge from the collection on the fringe. While many types of query are possible, four basic ones are of importance. Figure 43 shows the original graph. The Breadth First Search (BFS) takes the oldest of those available – the one put on the tree first. This grows the tree outwards like a bush (Figure 44). The dashed arrows indicate graph edges that were rejected by the Fringe Algorithm.

24

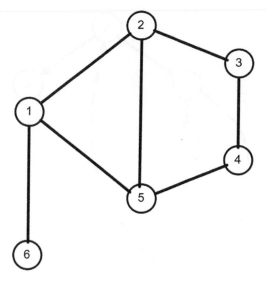

Figure 43.   The original graph.

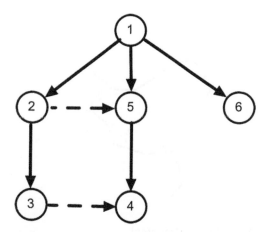

Figure 44.   Breadth first search.

The Depth First Search (DFS) (Figure 45) does the opposite: it takes the youngest first – the edge most recently added. It grows the tree downwards, like a root.

The Minimum Spanning Tree (MST) is used on a weighted graph, where the edges have 'lengths'. The fringe is searched for the shortest edge available at each step. Thus it will generate a tree with the shortest total length of edges. In a geographic context this could be imagined as connecting a series of towns – nodes – with the shortest possible total length of roads. The MST is often of use to us, as it is a subset of the Delaunay triangulation.

The Shortest Path (SP) is similar to the minimum spanning tree, but in this case a particular root must be specified at the beginning, and all distances will be measured back to this root. Thus the 'length' of a proposed edge is its own length – plus the distance from its parent all the way back to the root: this had been calculated in an earlier step. In a geographic context this could be imagined as deciding the delivery route from the root 'factory' to each of the destination locations.

These algorithms may be applied to graphs that are not planar, and to directed graphs – but in this second case a node may be reachable in one direction but not in the other.

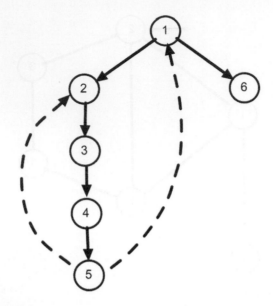

Figure 45.   Depth-first search.

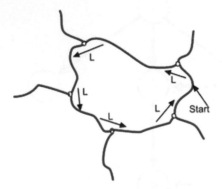

1: BFS graph traversal
2: Left child before right (needs node arc-order)
3: Closure on tree = polygon closure

Figure 46.   Polygon detection.

Polygon detection

Looking at the result of the breadth first traversal, an obvious geographic application comes to mind: if we are attempting to digitise some polygons on a map, and we have entered the nodes and edges, this traversal should be able to detect completed polygons whenever the algorithm attempts to close the loop. (Note that this only works with breadth-first traversal: depth-first will usually close the loop around multiple polygons.)

This approach is almost true but with one constraint: we need to be able to select the 'next' edge around each node – in clockwise order in this example (Figure 46). If we can do this then polygon construction is straightforward. However, standard graph theory has no mechanism for selecting the priority of edges around a node – we must add this ourselves.

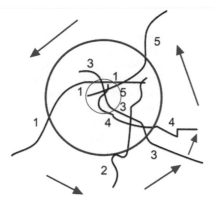

Topology needs the (CCW) order around each node

But this depends on the scale!

Figure 47.   Node detection.

This is not the only problem. In many cases nodes are detected at the intersections of digitised edges – but multiple edges will not usually intersect at precisely the same location. Our objective is to determine the order of edges around the possible node – but even the order is not obvious, as it depends on the scale and all the possible intersections or near-intersections of the digitised edges. In Figure 47 the intention is to have edges 1 to 5 in an anticlockwise order around the node, as shown by the big circle. However, if a smaller circle was chosen an order of $1 - 4 - 3 - 5 - 1 - 3$ would have been detected. This has been a particularly difficult job in GIS, and illustrates the fundamental weakness of a line segment model of space. We will look at other alternative spatial models later on. However, if we can determine the edge order around a vertex then polygon detection is straightforward.

Flow analysis

Any of the graph traversal techniques described above for undirected graphs produces a spanning tree – that is, a tree structure that visits all the nodes, starting at the root. So if we wish to examine the flow in a graph from the source (the root) to the sink (any other defined node) the traversal will give one possible path. However, our flow analysis assumes that each edge in our graph has a maximum capacity – perhaps the diameter of the pipe. So our original path will have a maximum flow defined by the lowest capacity of its various segments – its bottleneck. Other paths through the graph may be able to add to the total flow between the source and the sink. The algorithm to find the maximum flow therefore consists of performing a graph traversal from the source that includes the sink on one of its branches. The flow in this path is calculated as that of its bottleneck segment, and this value is subtracted from all the segments of the path between the source and the sink. (Any path segment whose capacity is that of the calculated flow in the path will now have an available capacity of zero – effectively removing that edge from the graph of available segments.)

This process is repeated and a new path found between the source and the sink, if possible. The flow in this path is again subtracted from its segments, and another path searched for – until on the final attempt no suitable path is found. This algorithm will find a maximum flow between the source and the sink – this will give a value for the maximum flow, but there may be other equally suitable combinations of paths.

This completes the algorithm – known as the Ford-Fulkerson method – with one exception: it is possible that a selected path flowing in one direction through a segment may in fact block the maximum solution, which would have a flow in the reverse direction in the same segment. To handle this possible situation we consider our undirected graph to be composed of two directed edges in opposite directions, and when we subtract our path flow from one direction we add it to the other – allowing the traversal algorithm to add flow in the reverse direction (which is equivalent to reducing it in the forward direction). This algorithm behaves well for simple flow values – such as integers – but may not be effective for fractional or decimal values.

The method may be extended simply when we have multiple sources, or multiple sinks, each with specified capacities, such as multiple electricity generators or multiple users. To handle this we create an 'infinite source' or an 'infinite sink' and connect these to the capacity-limited sources or sinks – but these new edges have flow capacities matching the specified limited capacities of the source or sink nodes. Thus, when the flow analysis is performed, no more than the limit may be assigned to these nodes.

Hopefully you can see that the straightforward graph algorithms we have described may have considerable utility in solving geographical problems, including route selection and maximum capacity flows.

In many mapping applications we first break the area into triangles with vertices at the data points and then need to process each of the vertices/edges/triangles once only – for example to draw the complete map. This required a sorting order for the triangles, so that each subsequent triangle is adjacent to one previously processed: a 'top to bottom' order – but points in 2D (or 3D) have no intrinsic ordering. Nevertheless, once they are triangulated it is possible to use the CCW test to identify each triangle's orientation with respect to some exterior point – the 'North Pole' – and use this to define a partial ordering – a graph traversal – that visits each triangle – or each edge, or each vertex – once and once only. This is described further in the section on scanned map processing.

FURTHER READING

Euler developed a solution to the Konigsberg bridge problem in 1735, by eliminating the irrelevant portions of the question (such as coordinates) and concentrating on the connectivity (landmasses and bridges). This led directly to the development of graph theory, and topology, as we know them today. For further history see Stander (1986).

In the 1950s, with the development of modern computers and telephone networks, several researchers worked on algorithms for enumerating all nodes and edges in a graph – often representing the connectivity of telephones and cables. This is usually called 'traversal', and is achieved by forming a spanning subtree of a graph – 'spanning' because it connects all nodes, and 'tree' or 'subtree' because of its tree-like form, with no closed loops. Kruskal (1956), Prim (1957) and Dijkstra (1959) developed algorithms still used today.

Shortly afterwards it was found that multiple calls to these algorithms could be used to model the flow in graph networks if the capacity of each link is known. The basic reference is Ford and Fulkerson (1962).

1.3 DOMINANCE

1.3.1 *Dominance, Voronoi, circle tests, parabolic uplifting*

Having dealt with Cartesian coordinates, and largely replaced them with local barycentric coordinates, we turn to questions concerning the spatial relationships between objects – initially points. Attempts to sort spatial objects by their X and Y (and Z) coordinates have been found to be problematical or else rather counter-intuitive, so we will attempt another approach: dominance.

The dominance region for a point (or, one might imagine, for a wild animal in its home) is the region in which it is more powerful than its neighbours: the further away it is from its home

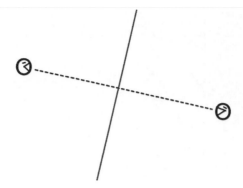

Figure 48.   Two-point dominance.

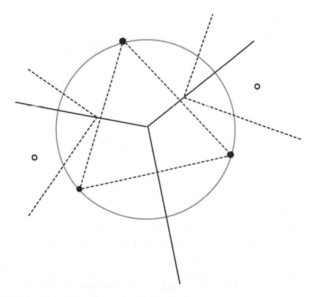

Figure 49.   A circumcircle, Voronoi node, and additional points.

base the more insecure it is, until it reaches its territorial boundary with a neighbour – if they are equally powerful this boundary will be equidistant between the two homes, thus dividing the map into two half-spaces (Figure 48) which we will later see to be their Voronoi regions. A straight line connecting the points is the dual of this Voronoi edge, and is part of what is called the Delaunay triangulation. For this approach to be meaningful we need to have some definition of equidistance, or distance: for most of our work this will be provided by Euclidean space, on a two-manifold – the surface of a sphere for example.

When we have three points we can show that, as each pair of points has an equidistant boundary half way between them, the intersection of any two of these boundaries (or all three) meet at a point equidistant between the three points: this is the centre of a circle (known as the circumcircle of the triangle – see Figure 49). A triangle is Delaunay if its circumcircle is empty, and the Delaunay triangulation of a set of points has all the circumcircles empty. (A sort-of exception is if four or more points lie on the same circle, as for a square, but this is easily handled by splitting the square into two triangles, each with the same coordinates for the centres of their circumcircles – their circumcentres.)

Figure 50.   VD/DT on a balloon.

The dominance regions (Voronoi cells) for the three points are thus convex polygons (with infinite external boundaries in this case). If an additional point is added to Figure 49 , and if it falls outside the circumcircle, then a new circumcentre is formed for the new triangle, as shown. (Note: in fact, two new triangles and circumcentres are formed for every point added, but these are hidden from view 'at infinity' on our sheet of paper. Draw the diagram on a balloon and this will become clear!)

By hand, it is easy to show that the Voronoi diagram (and hence the Delaunay triangulation) may be constructed locally, without having to change everything: just draw a few points on a piece of paper (or a balloon) and sketch the bisectors between any one point and some neighbouring ones – you will end up with a convex polygon around your chosen point (Figure 50) plus a few extra trial bisectors that weren't really needed. Once you have the Voronoi cells the generating points can be connected if they have a common boundary – and you have the Delaunay triangulation.

This is the best method for hand construction: in the computer it is better to work with the Delaunay triangulation and test if the new point V_P falls within the original circumcircle of the triangle. If it does, then the triangle is no longer Delaunay, and it must be split into three with V_P as a vertex of each, and the results adjusted, as will be described below.

Testing if V_P falls within the circumcircle is performed by a variant of the 3D CCW ('InCircle'), using the 'lift-up transform', where we let $z = x^2 + y^2$. All points then fall on a paraboloid of revolution (Figure 51).

If our point V_P has coordinates (x_P, y_P, z_P) – or $(x_P, y_P, x_P^2 + y_P^2)$, and the vertices V_1, V_2 and V_3 of our triangle are treated similarly, then:

$$\text{InCircle} = 3\text{DCCW} = \begin{vmatrix} x_P & y_P & z_P & 1 \\ x_1 & y_1 & z_1 & 1 \\ x_2 & y_2 & z_2 & 1 \\ x_3 & y_3 & z_3 & 1 \end{vmatrix} = 6 \times \text{the volume of a tetrahedron.}$$

Figure 51 shows a 2D triangulation (with positive CCWs – anticlockwise triangles). Our 2D circle, when lifted up, can be thought of as a plane in 3D: this slices the paraboloid such that the projection of the intersection is a circle on the 2D plane touching the three triangle vertices. The

30

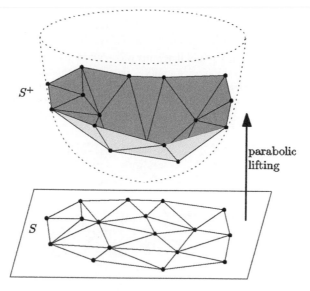

Figure 51. InCircle test.

value of InCircle is (6 times) the volume of the tetrahedron formed by the triangle vertices V_1, V_2 V_3 and V_P: if V_P is 'above' the triangle then the result is positive, if it is below it then it is negative, and if it is on the same plane it is zero. ('Above' can be taken as in the positive direction of the vector normal of (V_1, V_2, V_3). Figure 51 shows the vector normal of the triangle.) If V_P is outside the circle it is above the plane, and the volume ('InCircle', often called 'H') is positive. If $H = 0$ then V_P lies on the plane (and on the circle) and if H is negative then V_P falls below the plane and V_P is inside the circle.

Showing this is true can be done as follows: the equation of a circle can be written as: $x^2 + y^2 + ax + by + c = 0$. If we create a third dimension where $z = x^2 + y^2$ (giving a paraboloid) and substitute, then the equation of the circle can be expressed as a linear equation – a plane – in 3D: $z + ax + by + c = 0$. This plane cuts the paraboloid in an ellipse, which projects onto the 2D plane as a circle – the circumcircle of the triangle.

We can perhaps summarize by quoting O'Rourke (1998), as saying: 'In a sense a Voronoi diagram records everything one would ever want to know about proximity to a set of points (or more general objects).' Here are some of its properties:

a) Each Voronoi region is convex.
b) A Voronoi vertex is the centre of a circumcircle of three points.
c) The circumcircles of a DT are empty.
d) The nearest pairs of points form edges on the DT.
e) Every Delaunay edge has an empty circle.
f) The VD is unique for a given set of points. The DT is also unique if points are not co-circular.
g) In the DT lift-up: it is the only triangulation where the lift-up is convex-downwards.
h) The DT maximises the minimum angles of the triangles.
i) The largest empty circle is centred on a Voronoi vertex.

And (2D or 3D) CCW gives us the tool to manage the Voronoi diagram without further need to mess about with coordinates directly.

FURTHER READING

The biggest initial problem with Voronoi diagrams was – what to call them. The concept – that a region of space 'belongs' to the closest entity – dates to prehistory, and actually predates man: think of lions' dens, nesting sticklebacks, foraging seagulls. But many disciplines have discovered/rediscovered/used the concept. Figure 52, a 4th century Byzantine mosaic from the basilica of Poreč, Croatia, suggests such a diagram. (Score one for artists.)

Figure 52.   Mosaic.

Descartes suggested such a partition of space with respect to the planets in 1644 (republished in: Descartes 1728). (Score one for early cosmology.)

Dr. John Snow (1855), the father of modern epidemiology, used the same principle to locate the source of bad water during a cholera outbreak in 1854, and demonstrate that it was due to contaminated water. (Score one for epidemiology.)

G.L. Dirichlet (1850) used these 2D and 3D cells in his study of quadratic forms. (Score one for mathematics: they are sometimes called Dirichlet tessellations.)

Georgy Voronoi (or Vorony) also studied quadratic forms, and extended the method to higher dimensions. His name is most often associated with these cells – see Voronoi (1907, 1908).

Boris Delaunay (or Delone) was a student of Voronoi and defined what is now called the Delaunay triangulation Delaunay (1928, 1934). (He did other things too: he died in 1980 at 90 years old, he last published in 1970, and he was still strenuously mountain climbing at 80!)

The three articles by Mike Shamos and collaborators – see Shamos and Hoey (1975), Shamos 1978, 1985) – form the main start of modern 'Computational Geometry', based on analysis of the efficiency of algorithms. For geometric problems this can be quite complicated – in this book we only consider a few results, without proof, but all modern work is based on this approach.

Graz University of Technology has produced several excellent computational geometers, in particular Franz Aurenhammer and Herbert Edelsbrunner. Aurenhammer's 1991 review of Voronoi diagrams is one of the most highly cited in the field, and his 2013 book is probably the most up to date survey at present – but not easy reading! Edelsbrunner is best known for 'simulation of simplicity' techniques, where algorithms that are only theoretically valid for data in 'general position' (usually when points are not collinear) may be made robust for all cases. He is also well known for his work on 'alpha shapes', for generating surfaces from 3D point clouds.

For the non-expert, O'Rourke's 1998 book, and that of de Berg et al. (2000) are probably the best introduction. O'Rourke's book gives working code for many algorithms, and de Berg's book gives excellent simple-to-follow descriptions of basic Voronoi, Delaunay and other computational geometry algorithms.

My own entry into the field came from my PhD work on mapping subsurface geology from drill-hole data. Contour mapping by constructing grids of interpolated values, and then drawing in the contours, was the method of choice in geology and geography in the 1970s. Starting from scratch, I reasoned that flat triangular plates connecting the data points would at least match the elevation at the data points – a point of view I have never changed! Given this, I had to define a data structure for inserting points into a triangulation, and an algorithm – and lastly a definition of what constituted a 'good' triangle. My 1978 symposium paper was reviewed by Mike Shamos, who introduced me to the Delaunay triangulation: I have never looked back! (See Gold (1978, 1988, 1989, 1991, 1992), Gold and Cormack (1987), Gold et al. (1977).)

1.3.2 *Incremental algorithm*

The InCircle test gives a simple method for finding if we have a valid Delaunay triangle, as this must have an empty circumcircle. If we have inserted a new point into a triangulation, splitting the old triangle into three after we have found the correct one, then the resulting triangles must be tested to see if the fourth point, from an adjacent triangle, is inside the circle or not: if it is inside then the triangle pair must be switched to produce two better (valid) triangles. This is the basic incremental algorithm (Figure 53).

It is always true that a convex quadrilateral formed by two triangles may be split so that the circumcircles of the two triangles are empty – or else, split the other way, they are not. In Figure 54, P_4 is outside the circumcircle of (P_1, P_2, P_3) – and similarly for P_1 with respect to (P_2, P_4, P_3). However, if the diagonal connected P_1 and P_4 instead, two different triangles would be formed, and P_3 would be inside the (much larger) circumcircle of (P_1, P_2, P_4). Similarly P_2 would be inside the large circumcircle formed by (P_1, P_4, P_3). (Note that the vertices are defined in anticlockwise order.)

Thus if the quadrilateral had been split vertically then the 'test' operation using triangle (P_1, P_2, P_4) would find that P_3 was inside the circumcircle (as the 3D InCircle test, described earlier, was negative), and the 'swap' operation would produce two valid Delaunay triangles (with empty circumcircles) as seen in Figure 54.

If a new point P was inserted in the figure, and one of the triangles split into three, then each new triangle edge, plus its matching triangle on the exterior would need to be tested, as it was possible that the two new triangles had other, further out, points within their circumcircles. Thus testing, and swapping, must continue until no further points within circumcircles are found: each swap adds another Delaunay neighbour to the newly inserted point P.

This is usually done with a 'stack' to keep track of which edges remain to be tested. It can be shown that, on the average, a point has six neighbours in a Delaunay triangulation, and so the total number of swaps is usually small. This shows one of the major advantages of the Delaunay triangulation: inserting (or deleting) a single point only makes local changes to the structure, meaning it can be updated on a point by point basis, and the order of the input points is unimportant. (This property can also be easily shown with the manual method: adding a new Voronoi cell on your balloon only affects a small number of neighbours.) We now have a valid Delaunay triangulation, and can

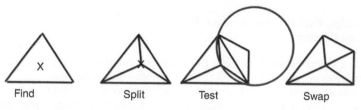

Figure 53.   Incremental algorithm.

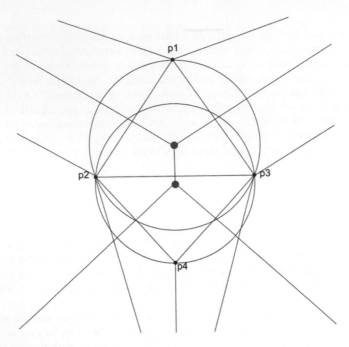

Figure 54.   Diagonal switching.

proceed to add the next point: this is called the Incremental Insertion Algorithm. While not the fastest algorithm, it is robust, fairly fast, and simple.

A very important consequence of this is that, since inserting or deleting a single point only changes things locally, the resulting triangulation is unique for that set of points. (A minor exception is when four or more points are co-circular – then the larger region may be split in various ways – as when P_4 is precisely on the circumcircle of (P_1, P_2, P_3).)

In summary, for the Voronoi diagram each cell contains all locations closer to a generating point GP than to any other; boundaries are equidistant to two GPs; and Voronoi nodes are equidistant to three GPs – these circles are empty. The Delaunay triangulation is the dual of the Voronoi diagram; the circumcircles of these triangles are empty; its construction is unique and updating is local – additional co-circular points may be handled by splitting the larger region into triangles with a common circumcentre.

The Incremental Insertion Algorithm may be summarized as:

Create a bounding triangle (and its 'back')
'Walk' the triangulation to locate new pt. P
Split this triangle into 3
'Push' the exterior 3 edges onto a stack
'Pop' an edge from the stack. Test if it is 'Delaunay'. (Its 2 triangles have empty circumcircles)
If yes: **Pop**.
If no: flip the edge, **Push** the exterior edges, **Pop**
Repeat until stack is empty.

The result is a Voronoi diagram and/or a Delaunay triangulation. For the Delaunay triangulation all the circumcircles are empty. Figure 55 shows a simple Voronoi diagram.

Figure 56 shows several empty circumcircles in a Delaunay triangulation.

Figure 57 shows several stages in the insertion of a new data point. In the top left diagram a point is inserted and the triangle split into three. In the next, the lower left edge is tested, and a vertex is

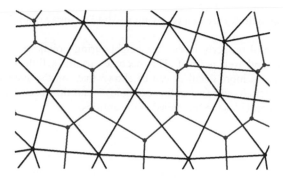

Figure 55.   The VD and DT.

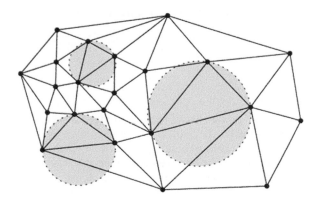

Figure 56.   Empty circumcircles.

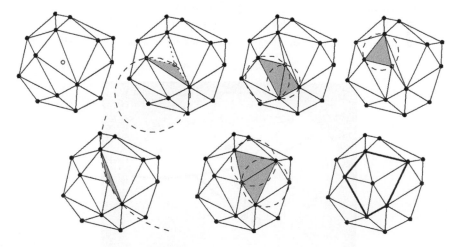

Figure 57.   The complete insertion operation.

found to fall inside the circumcircle. The diagonal is swapped, and the resulting pair of triangles
are locally correct. On the right hand side another of the original split triangles is tested, and found
to have an empty circumcircle. In the bottom left figure another of the original split triangles is
tested – it has a very large circumcircle, and several points fall inside it. The diagonal is swapped

and the remaining circumcircles are empty, giving a valid Delaunay triangulation around the new point.

One immediately useful property is that the neighbouring points to any particular point are immediately available, and well defined. This becomes useful a little later when we examine interpolation techniques in more detail. It can be shown that, on the average, each Voronoi cell has six neighbours (although only five in this example) – and thus each Delaunay vertex has six adjacent triangles. This provides the 'domains' of the entities.

FURTHER READING

Starting in the 1970s, various authors developed algorithms for the construction of the 2D DT/VD, with later emphasis on robustness and numerical stability (See Lloyd (1977), Green and Sibson (1978), Lee and Schachter (1980), Bowyer (1981), Watson (1981), Guibas and Stolfi (1985), Fortune (1987), Sugihara and Iri (1992), Fortune (1995), Seidel (1998), Sugihara et al. (2000)).

1.3.3 *The VD/DT on the sphere*

When constructing the VD/DT on the plane, we have to make a special case for points on the boundary (more precisely, on the convex hull) as the Voronoi edges extend to infinity. It is much easier to visualise it on the sphere, with no boundary conditions, where all boundary Voronoi edges meet at the back. This is quite simple, as three vertices on the sphere's surface form a triangle: the triangle is on a plane that cuts the sphere, the circumcircle is the intersection of the plane with the sphere, and the circumcentre is along the vector normal of the triangle, as shown in Figure 58.

A portion of the VD/DT on a sphere is shown in Figure 59.

In one project individual points, representing packets of water, were perturbed and the graph updated, as described later under 'Free-Lagrange flow modelling'.

As an aside, the VD on the sphere has a variety of geographic applications. For example:

"Ürümqi, in China, is the city that is furthest from the sea, in the whole world" – is this true?

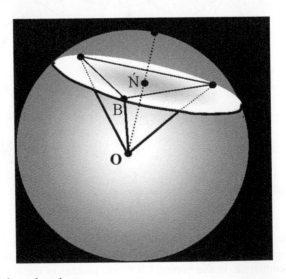

Figure 58.   Circumcircle on the sphere.

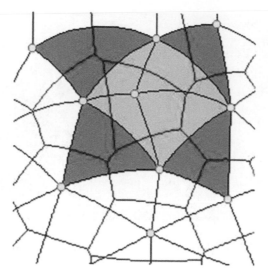

Figure 59.   Part of the VD/DT on the sphere.

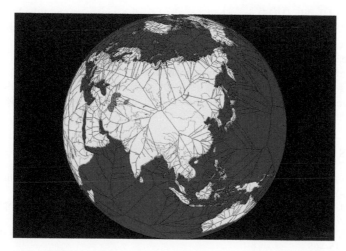

Figure 60.   The VD of coastlines.

Answer: calculate the Voronoi diagram, on the sphere, with the world's coastlines added as points. The Voronoi node with the largest radius is the furthest from the sea! (So it is true: Ürümqi is in the centre of the figure.) Figure 60 shows the Voronoi diagram, and Figure 61 shows the corresponding Delaunay triangulation: the largest triangles have the largest circumcircles. China is coloured yellow.

In another application, if we have a set of orientation data, as with geological strike-and-dip (orientation) observations of some exposed folded surface, we may well want to see how and whether they represent one group of orientations, or several – or even a band, as for observations along the curve of a folded formation. The orientations of the individual observations may be plotted, as unit vectors, upon the surface of a unit sphere or hemisphere, and the VD/DT drawn as described previously. The minimum spanning tree, described below, can be shown to be a subset of the edges of the DT, and this is then used to identify clusters of observations with a similar orientation, as will be discussed later under cluster analysis, and then LiDAR roof modelling.

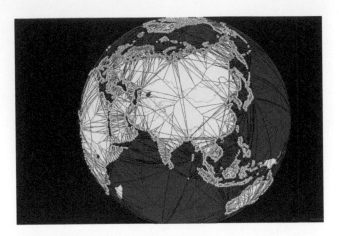

Figure 61.   The DT of the coastlines.

FURTHER READING

Augenbaum and Peskin (1985) described a method for the construction of the Voronoi mesh on the surface of a sphere. Of particular interest is the observation that the circumcircle is defined by the intersection of the plane of the Delaunay triangle with the sphere surface. Renka (1997) developed the code for an incremental algorithm to construct the VD and DT on the surface of a sphere. Mostafavi (2004) worked on constructing a kinetic Voronoi diagram on the surface of a sphere, for modelling 2D fluid flow of tides. Of particular interest is the implementation of Free-Lagrange simulation, bypassing the use of fixed intervals in space and time.

REFERENCES

Angel, E. (2006) *Interactive Computer Graphics: A Top-Down Approach with OpenGL*. 5th edition. Boston, MA, Addison-Wesley.

Augenbaum, J.M. & Peskin, C.S. (1985) On the construction of the Voronoi mesh on a sphere. *Journal of Computational Physics*, 59, 177–192.

Aurenhammer, F. (1991) Voronoi diagrams – A survey of a fundamental data structure. *ACM Computing Surveys*, 23, 345–405.

Aurenhammer, F. & Edelsbrunner, H. (1984) An optimal algorithm for constructing the weighted Voronoi diagram in the plane. *Pattern Recognition*, 17, 251–257.

Aurenhammer, F., Klein, R. & Lee, D.T. (2013) *Voronoi Diagrams and Delaunay Triangulations*. Singapore, World Scientific Publishing Company. Blinn, J.F. (1977) A homogeneous formulation for lines in 3 space. *ACM SIGGRAPH Computer Graphics*, 11, 237–241.

Blinn, J.F. (1977) A homogeneous formulation for lines in 3 space. *ACM SIGGRAPH Computer Graphics*, 11, 237–241.

Bowyer, A. (1981) Computing Dirichlet tessellations. *The Computer Journal*, 24, 162–166.

Coxeter, H.S.M. (1969) Barycentric coordinates. In: *Introduction to Geometry*. Vol. 13. New York, NY, Wiley.

de Berg, M., van Kreveld, M., Overmars, M. & Schwarzkopf, O. (2000) *Computational Geometry: Algorithms and Applications*. 2nd revised edition. Berlin, Springer-Verlag. pp. 147–163.

Delaunay, B. (1928) Sur la sphere vide. In: *Proceedings of the International Mathematical Congress held in Toronto, August 11–16*. Toronto, University of Toronto Press, 69–70.

Delaunay, B. (1934) Sur la sphère vide. *Bulletin of Academy of Sciences of the USSR*, 6, 793–800.

Descartes, R. (1728) *Le monde, ou Traité de la lumière*.

Dijkstra, E. (1959) A note on two problems in connection with graphs. *Numerische Mathematik*, 1, 269–271.

Dirichlet, G.L. (1850) Uber die Reduction der positeven quadratischen Formen mit drei unbestimmten ganzen Zahlen. *Journal fur die Reine und Angewandte Mathematik*, 40, 209–227.

38

Douglas, D. (1974). It makes me so CROSS. Unpublished manuscript from the Harvard Laboratory for Computer Graphics and Spatial Analysis. Reprinted in: Peuquet, D.J. & Marble, D.F. (eds.) *Introductory Readings in Geographic Information Systems*. London, Taylor & Francis. pp. 303–307.

Edelsbrunner, H. & Seidel, R. (1986) Voronoi diagrams and arrangements. *Discrete Computational Geometry*, 1, 25–44.

Faux, I.D. & Pratt, M.J. (1979) *Computational Geometry for Design and Manufacture*. Chichester, Ellis Horwood.

Foley, J.D., van Dam, A., Feiner, S.K. & Hughes, J.F. (1990) *Computer Graphics: Principles and Practice*. Reading, MA, Addison-Wesley.

Ford, L.R. & Fulkerson, D.R. (1962). *Flow in Networks*. Princeton, NJ, Princeton University Press.

Fortune, S. (1987). A sweepline algorithm for Voronoi diagrams. *Algorithmica*, 2, 153–174.

Fortune, S. (1995). Numerical stability of algorithms for 2-D Delaunay triangulations. *International Journal of Computational Geometry and Applications*, 5, 193–213.

Gold, C.M. (1978) The practical generation and use of geographic triangular element data structures. In: Dutton, G. (ed.) *Proceedings First International Advanced Study Symposium on Topological Data Structures for Geographic Information Systems. Harvard Papers on Geographic Information Systems, Volume 5 – Data Structures: Surficial and Multi-Dimensional*. Cambridge, MA, Laboratory for Computer Graphics and Spatial Analysis, Harvard University. pp. 1–18.

Gold, C.M. (1988) PAN graphs: An aid to G.I.S. analysis. *International Journal of Geographical Information Systems*, 2, 29–42.

Gold, C.M. (1989) Spatial adjacency – A general approach. In: *Proceedings, Auto-Carto*. Vol. 9. pp. 298–312.

Gold, C.M. (1991) Problems with handling spatial data – The Voronoi approach. *CISM Journal*, 45, 65–80.

Gold, C.M. (1992) The meaning of 'neighbour'. In: *Lecture Notes in Computing Science 639: Theories and Methods of Spatio-Temporal Reasoning in Geographic Space*. Berlin, Springer-Verlag. pp. 220–235.

Gold, C.M. & Cormack, S. (1987) Spatially ordered networks and topographic reconstructions. *International Journal of Geographical Information Systems*, 1, 137–148.

Gold, C.M., Charters, T.D. & Ramsden, J. (1977) Automated contour mapping using triangular element data structures and an interpolant over each triangular domain. In: George, J. (ed.) *Proceedings Siggraph '77. Computer Graphics*. Vol. 11. pp. 170–175.

Gold, C.M., Chau, M., Dzieszko, M. & Goralski, R. (2004) 3D geographic visualization: The marine GIS. In: Fisher, P.F. (ed.) *Developments in Spatial Data Handling – 11th International Symposium on Spatial Data Handling*. Berlin, Springer. pp. 17–28.

Green, P.J. & Sibson, R. (1978) Computing Dirichlet tessellations in the plane. *The Computer Journal*, 21, 168–173.

Guibas, L. & Stolfi, J. (1985) Primitives for the manipulation of general subdivisions and the computation of Voronoi diagrams. *ACM Transactions on Graphics*, 4, 74–123.

Huang, C.-W. & Shih, T.-Y. (1997) On the complexity of point-in-polygon. *Computers & Geosciences*, 23, 109–118.

Kruskal, J.B. (1956) On the shortest spanning subtree of a graph and the travelling salesman problem. *American Mathematical Society*, 7, 48–50.

Lee, D.T. & Schachter, B.J. (1980) Two algorithms for constructing the Delaunay triangulation. *International Journal of Computer and Information Sciences*, 9, 219–242.

Lloyd, E.L. (1977) On triangulations of a set of points in the plane. In: *Proceedings of the 18th Annual IEEE Conference on the Foundations of Computer Science*. pp. 228–240.

Mostafavi, M.A. & Gold, C.M. (2004) A global spatial data structure for marine simulation. *International Journal of Geographical Information Science*, 18, 211–227.

O'Rourke, J. (1998) *Computational Geometry in C*. 2nd edition. Cambridge, Cambridge University Press.

Phong, B. (1975) Illumination for computer generated pictures. *Communications of the ACM*, 18, 311–317.

Preparata, F.P. & Shamos, M.I. (1985) *Computational Geometry – An Introduction*. New York, NY, Springer-Verlag.

Prim, R. (1957) Shortest connection networks and some generalizations. *Bell Systems Technical Journal*, 36, 1389–1401.

Renka, R.J. (1997) Delaunay triangulation and Voronoi diagram on the surface of a sphere. *TOMS*, 23, 416–434.

Saalfeld, A. (1987) It doesn't make me nearly as CROSS: Some advantages of the point-vector representation of line segments in automated cartography. *International Journal of Geographical Information Systems*, 1, 379–386.

Sedgewick, R. (1983) *Algorithms*. Reading, MA, Addison-Wesley.

Seidel, R. (1998) The nature and meaning of perturbations in geometric computing. *Discrete and Computational Geometry*, 19, 1–17.

Shamos, M.I. (1978) *Computational Geometry*. PhD Thesis. Ann Arbor, MI, Yale University. University Microfilms.

Shamos, M.I. & Hoey, D.J. (1975) Closest-point problems. In: *Proceedings of the Sixteenth IEEE Symposium on Foundations of Computer Science*. pp. 151–162.

Shewchuk, J.R. (1997) Adaptive precision floating-point arithmetic and fast robust geometric predicates. *Discrete and Computational Geometry*, 18, 305–363.

Snow, J. (1855) 'Dr. Snow's report' in: Report on the cholera outbreak in the Parish of St. James, Westminster, during the autumn of 1854 presented to the Vestry by the Cholera Inquiry Committee. London, J. Churchill. pp. 97–120.

Stander, D. (1986) The Euler Formula – Its history, applications and teaching. *Teaching Mathematics and Its Applications*, 5, 112–126.

Sugihara, K. & Iri, M. (1992) Construction of the Voronoi diagram for 'one million' generators in single-precision arithmetic. *Proceedings of the IEEE*, 80, 1471–1484.

Sugihara, K., Iri, M., Inagaki, H. & Imai, T. (2000) Topology-oriented implementation – An approach to robust geometric algorithms. *Algorithmica*, 27, 5–20.

Voronoi, G. (1907) Nouvelles applications des parametres continus a la theorie des formes quadratiques Premier Memoire: Sur quelques proprietes des formes quadratiques positives parfaits. *Journal für die Reine und Angewandte Mathematik*, 133, 97–178.

Voronoi, G. (1908) Nouvelles applications des paramètres continus à la théorie des formes quadratiques. *Journal für die reine und angewandte Mathematik*, 134, 198–287.

Watson, D.F. (1981) Computing the n-dimensional Delaunay tessellation with application to Voronoi polytopes. *The Computer Journal*, 24, 167–172.

Wolberg, G. (1994) *Digital Image Warping*. Los Alamitos, CA, IEEE Computer Society Press.

Spatial Context: An Introduction to Fundamental Computer Algorithms for Spatial Analysis – Gold
© 2016 Taylor & Francis Group, London, ISBN 978-1-138-02963-7

Chapter 2

Models of space

2.1 MODELS OF SPACE: INTRODUCTION

Perhaps one of the least understood aspects of spatial analysis is the fact that all of our algorithms and analyses presuppose one or another of a variety of models of space – perhaps 'conceptual models of space' is more appropriate.

We were raised in an environment where an X Y system of coordinates, and squared graph paper, were the foundations of our attempts to make sense of the objects within it. It is very hard to shake off this perception, and consider other ways of thinking about even two dimensional problems – although we can probably accept that the number of dimensions is a basic property of the space we wish to manipulate. Nevertheless, the most famous book of geometry ever written, by Euclid, was written long before the development of coordinate systems.

The first section of this book attempts to show some of the limitations of Cartesian coordinates within a computer environment – let alone within the real world we wish to emulate. Vectors are pre-coordinates, and can be demonstrated as well in the sand as on the screen. The fundamental predicate CCW and its extensions were derived from vectors, and are used to replace coordinates with more 'primitive' queries, such as sidedness, that are usually more easily handled by human intuition. The dominance regions for a pair of points are easily handled with a compass and ruler, and the same is true for three points or more in the Voronoi diagram: the equivalent derivations from the point coordinates uses InCircle – derived from CCW 3D, in turn extended from the 2D CCW predicate. So (except for our graphical output when we return to the computer world of coordinates again) we are using the computer to simulate Pythagoras's drawing in the sand.

The way we think of space varies considerably depending on the application, the type of data and our own presuppositions. One well-accepted distinction is between a set of spatially-located 'objects' such as buildings and a continuously-varying value – a 'field' – such as temperature. Another useful distinction concerns our navigation within this world: are we a 'plane' flying above, or a 'boat' acting within this space and interacting with it – this would mostly, but not necessarily, apply to the 'object' model.

A 'field' is continuous, but not necessarily smooth: it could consist of a tessellation with constant values in each cell – 'blocks'. An 'object' model would have empty space between at least some of the tiles, and does not necessarily have any obvious spatial relationships between them.

A famous debate between Newton and Leibnitz concerned whether space was 'real' and could be measured (Newton) or only existed as the spatial relationships between objects (Leibnitz). The first led to Cartesian coordinate systems and raster spatial models, and the second led to graph theory and networks. (We can of course have 2D and 3D Cartesian space – and other dimensions – but other metric spaces are possible, including 'city block metric' where distance must be measured along the axes only.) The early chapters of this book are an attempt to move from Newton towards Leibnitz, at least as far as computer-based spatial analysis is concerned.

We may also have spatial embedding, when a lower dimension structure may be embedded in a higher dimension – for example a terrain surface embedded in 3D space, a road embedded in 2D space or a node embedded in a 1D road. (All of these examples imply that the higher-dimensional space is not necessarily fully occupied: that is, an arbitrary query at any location in the bounding space will not always return a value.)

The simplest form of spatial relationship between a pair of objects is adjacency: if their boundaries touch (presupposing that only one object may occupy any particular location) then a relationship – an edge of a graph – may be recorded. In a field model the whole space – within some bounded region – is filled and all objects (tiles) have recorded neighbours. In an object model this will, at most, be only partially true, meaning that navigation within that model cannot be guaranteed.

The Voronoi spatial model attempts to remedy that deficiency: objects are expanded until all boundaries meet, 'holes' in the space are filled in, and a complete set of spatial adjacency relationships are obtained: the object model becomes a field. As with fields, a response is available at any spatial location and, as with the object model, the 'name' of the object and its spatial properties may be obtained directly (unlike traditional raster fields, where the object must be reconstructed from grid cell attributes). Only one change needs to be considered: instead of obtaining 'the object at this location is ...' we get 'the object closest to this location is ...'. As we will see, this is often what we wanted to ask in the first place.

In its simplest form, an object (often a point) dominates all spatial locations closer to it than to any other point: this is its dominance region – think perhaps of predators who protect their territory with greater ferocity closer to their lair: there will be a boundary between two territories at an equal distance from the lairs. This boundary is the perpendicular bisector of the line between these generators. For three generators – and three boundaries – there is a single location equidistant between the three, at the centre of a circle touching the three generators, where the three boundaries meet. (For collinear generators and boundary cases it is often easier to place them on a sphere rather than a plane. In 3D the separating boundaries are planes, with four meeting at the circumcentre.) Adding additional generators produces a complete tessellation of the space, termed a Voronoi diagram (VD). The collection of the edges connecting generators sharing boundaries is termed the Delaunay triangulation (DT). (In the plane the DT forms a set of triangular regions except where four or more generators are co-circular, in which case that region may be arbitrarily subdivided into triangles without losing any of the useful properties.)

We are now free to consider spatial models that are not, of themselves, based on a global coordinate system. But let's start at the beginning.

2.2 THE COSMIC SPATIAL MODEL

Over the years GIS has developed various models of space, depending on the application and the particular computer algorithm involved. This has caused a variety of difficulties, not least when transferring data from one spatial model to another – for example between contour maps and triangulated irregular terrain models. It is time therefore to step back and look at the 'big picture' again, to see to what extent we can have a single coherent underlying model of the space we are working in.

So: what is space, and why do we care? We want to describe objects in it, and, more importantly, we want to describe their relationships to other – adjacent – objects. (We might call this the 'context' of the object.) So what is the 'big picture'? The earth? The Galaxy? Perhaps the universe! So, to do some 'blue sky' thinking – but this is really 'beyond the sky' – we will start by asking: what do we know about the universe?

Modern cosmology provides some intriguing thoughts: work by Icke and van de Weijgaert, looking at the recent to 2DF Galaxy red shift survey – the attempt to map a slice of the universe as far out as is possible – as well as other recent surveys suggests that the distribution of matter (galaxies) is concentrated in clusters, strings and even slabs. In between there are voids, where there are rather few galaxies observed. The intriguing question is: what is inside these voids? There are a few stray galaxies, and presumably dark matter, about which we do not know very much, and – gravitational force fields, generated by the individual galaxies. That's about it.

How did the pattern get that way? Presumably it was more homogeneous earlier on.

Voids expand. That is, any initial little voids will tend to grow bigger, like bubbles. Until they push the galaxies to the edges, into slabs. Why?

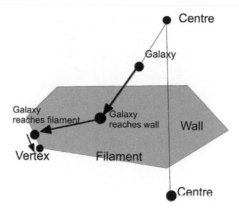

Figure 62.   Galaxy migration.

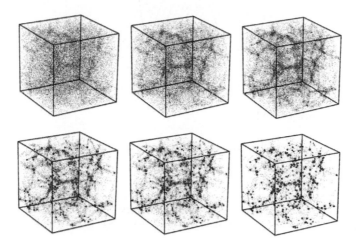

Figure 63.   Simulation of galaxy migration.

Well, groups of galaxies exert gravitational pull. Voids don't. So any stray galaxies, lost and lonely in the void, get attracted to their neighbours along the edges (3D faces, or slabs) until all that is left in the void is: gravity fields (Figure 62).

And in the slabs, galaxies wander to where two slabs meet – edges, with even more galaxies, and even more gravitational pull. And along edges they wander to nodes where edges meet, and have even higher density. This process isn't finished yet, but we can see the pattern.

"When a galaxy tries to enter an adjacent cell, the velocity component perpendicular to the cell wall disappears. Thereafter, the galaxy continues to move within the wall, until it tries to enter the next cell; it then loses its velocity component towards that cell, so that the galaxy continues along a filament. Finally, it comes to rest in a node, as soon as it tries to enter a fourth neighbouring void." Rien van de Weijgaert (2007).

Soap bubbles do almost the same thing, although the mechanism is different. Get soap bubbles on a flat surface and watch liquid flow from the centre of the skin towards the junction with another bubble, and then along the edge towards a node.

How to prove it? We can simulate it – quite successfully, with a big computer. And the result is: a Voronoi diagram! (Well, sort of – at least a good approximation.) Figure 63 shows the result of one simulation.

So: Space is one great big field (in the geographic sense) – with objects embedded in it. And these objects generate local gravitational sub-fields. And the local field extent can be estimated by

43

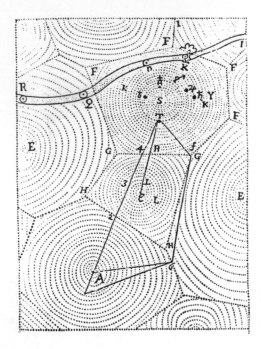

Figure 64.   Descartes' model of stellar forces.

the Voronoi diagram. As far back as Descartes this idea was being envisaged for the influences of stars (Figure 64).

And as Einstein (1952) stated:

> "I wished to show that space-time is not necessarily something to which one can ascribe a separate existence, independently of the actual objects of physical reality. Physical objects are not in space, but these objects are spatially extended. In this way the concept of 'empty space' loses its meaning."

Pairs of these generators (galaxies) have opposing 'pull'. At the mid-point the pull is equal, and we have zero gravity. Closer to either side an object will be drawn to the nearest generator. This is a simple dominance diagram (Figure 48). We could also imagine two predators defending their territory.

Three points have three dominance pairs – with boundaries that meet at the centre of the circumcircle of their connecting triangle (Figure 49). (In 3D we need 4 points, a tetrahedron and a circumsphere.)

Add more points and we get more 2D triangles and circumcircles (Figure 56) (or 3D tetrahedra and circumspheres). We see this effect in many places in nature, with a wide variety of generating mechanisms – e.g. Figure 65.

So much for individual objects/galaxies. What about the cosmic voids?

In our cosmic model, voids emptied their galaxies towards their own boundaries.

The voids are empty – of galaxies. But they are full – of the force fields of their boundary galaxies.

Suppose we colour galaxies migrating from one void to that void's boundaries. Then in the end we will have each void with different coloured force fields. We would have coloured our polygons/polyhedra.

Figure 66 is a sketch of galaxies along a slab boundary between two voids. The latest galaxies to arrive are along the edges of the slab. These galaxies didn't just rush through the boundary – they merged. They stayed on their own side – the boundary was impermeable. The same with the adjacent void: the boundary was two-sided.

Figure 67 is a sketch if we ignore the galaxies in the centre.

44

Figure 65.   Giraffe skin.

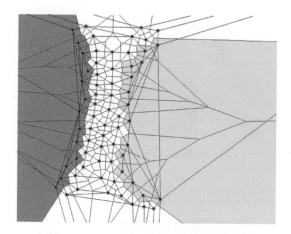

Figure 66.   Galaxies along a 'slab'.

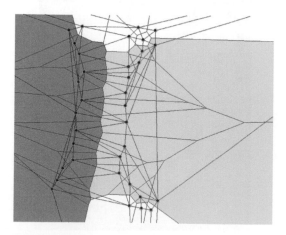

Figure 67.   Boundaries of a 'slab'.

We have two sides to each boundary. And each side has an associated void and also an adjacent side of another void. This gives us our dual relationship – our Delaunay edge. Each half-boundary has a fixed dual half-edge.

The Dual is the Context! It tells how voids relate to other voids. We can navigate through Space this way.

Thus individual objects generate local sub-fields. Voids empty themselves towards their boundaries; boundaries become impermeable, two-sided; voids are occupied by the force-fields of their own boundaries; and duality ties together the half-boundaries and voids.

FURTHER READING

Most ancient civilizations studied the movement of the stars, and even developed mechanisms to predict them – such as the unique Antikythera Mechanism, from a 1st century BC shipwreck, as seen in Figure 68 as recovered, and a computer simulation (Figure 69).

Figure 68.   The Antikythera mechanism (photo by van de Weijgaert).

Figure 69.   Modern simulation of the mechanism (by permission, Mogi Vicentini, 2007).

Much later – around 1633 AD – Descartes sketched an outline of the relative influence of the stars.

Leibniz and Newton also discussed the form of space: in very simple terms, Newton thought of coordinates in a Cartesian (from Descartes) system (the absolute theory of space), while Leibniz thought in terms of the distances between objects (the relational approach) – see Clarke (1717). For our purposes, we may pick up on Einstein's comment, mentioned previously, that physical objects are not in space, but are spatially extended – pretty much a Voronoi diagram!

Since Newton, Descartes and Clarke thought in terms of absolute coordinates (think grid squares) and Leibniz thought in terms of relationships (think graphs, triangles) we may tentatively think that the Voronoi diagram integrates the properties of both. Einstein (1952) thought of space as an expansion of the objects within it.

More recently, astronomers, astrophysicists and cosmologists, armed with the most recent galactic surveys, have started to look at the large-scale structure of the universe, in terms of the distribution of galaxies. Examples are Yoshioka and Ikeuchi (1989), Coles (1990), Ikeuchi and Turner (1991) and Zaninetti (1991).

In particular, Icke, and then his student van de Weijgaert (who also looked at the Antikythera mechanism), articulated the expanding void model which we use here: for example Icke and van de Weijgaert (1991), and van de Weijgaert (1994, 2007). This perhaps gives a touch of respectability to the Cosmic Spatial Model which we develop in this book!

2.3   BACK TO EARTH

2.3.1   *First there were fields*

Returning to Earth, we would like to examine what 'space' is – not just in the computer but in reality. Perhaps the best way to start is to think of it as a 'field', as in physics, where magnetic or gravitational fields are in force. This space is continuous, and unbounded. We are all familiar with magnetic fields, from high school physics, and this idea will give us the basic ideas that we want to use – although cosmic gravitational fields are actually better, because magnetic fields have north and south poles, which confuse the issue for us at present.

We have said nothing about dimensionality as yet, but our ideas should be applicable in one, two, or three dimensions. In two dimensions this makes us think of a continuous surface, perhaps with an 'elevation' representing the strength of the magnetic field. In three dimensions the same principle holds, and in one-dimension we can imagine forces along a line or curve.

2.3.2   *Then there were things*

If this field is to be of interest to us, we need to embed objects – generators – within it. We would then like to examine their relationships, or interaction. Newton and Leibniz had a long discussion about this, while considering the interaction of the sun and the planets. Newton emphasised the existence of a global coordinate system, which in the end led to the global Cartesian coordinates that we use today. Leibniz emphasised the relationships between individual pairs of objects, in particular the distance between them, which led us to start thinking about networks and graphs.

If we embed a pair of points in this space and consider them to be generators of force fields, then the forces will cancel each other out precisely at the perpendicular bisector of the line between them – giving us the simple dominance diagram. If we embed three points within our field, assuming they are not all in a line, then these perpendicular bisectors between the three pairs of points meet at a common point – the centre of the circumcircle of the triangle formed by these three points. This gives us one way to look at the construction of the Voronoi diagram.

2.3.3   *Then there were connected things*

Points may be embedded in a subspace of the space field: a plane or 2D surface within our 3D world, such as a topographic surface; a 1D line or curve within a 2D surface; or a point within the locus of a curve. Points sufficiently close together may become complex objects that may be referenced as a group: later discussion of the crust and skeleton will emphasize this. The crust represents the connectedness of these points (the skin), the skeleton the relationships with other objects or portions of the same curve – think of a human skeleton. A curve is the locus of a moving point – or of its multiple potential locations. The crust and the curve are the discrete or continuous representations of the same object – the crust could be considered the sampling of the curve.

The same applies for the embedding of a 2D surface in 3D space: originally the surface was continuous, but then it was discretely sampled. 'Above' and 'below' may be specified as with 'left' and 'right' in 2D, but using the 3D CCW test instead of 2D. A 1D line may be embedded in this surface, with the same rules – and a 0D point may be embedded within the 1D line.

2.3.4 *Then there were continuous things*

However, in the real world, at a human scale, objects appear continuous, not as a series of samples. For a line in 2D space, for example, we cannot expect to be able to penetrate between the samples – it has a sidedness, an impenetrability. This can be simulated in a crust as a specification of the sample spacing with respect to the distance to the nearest exterior object.

If we have sidedness then we have sides and, for closed curves, an 'inside' and an 'outside'. This partitions our global space field into sub-fields – polygons. These may be discrete objects, such as the outline of a building, or they may be a coverage of the space field – a thematic map.

In all of these cases the skeleton expresses the relationships between the various objects – if they are adjacent (in the sense that there is nothing in between) then their force fields 'meet' (cancel each other out) at their common boundary. These boundaries have a sidedness: a new point is closer to one object than to another.

2.3.5 *Then there were sidedness and direction*

On this basis all curves, and skeletons, have a direction – but a direction associated with one side of the curve, not the whole: the two sides have opposite orientations. When a crust can no longer be penetrated we must be able to specify which side a new point is on: the unspecified directionality of the crust edge must be split into two half edges. If we assume as a convention that the skeleton boundaries of a point are oriented in anticlockwise order then this is true for the interiors of skeletons of complex objects. Thus curves are pairs of half-lines, as are skeletons, and so are the crusts – and also the remaining dual (Delaunay) edges, since they link with the crusts. Thus an appropriate data structure for discrete point-based entities is the single undirected edge. But once they are considered to be impenetrable then half-edge pairs are needed to clarify the sidedness. 3D polyhedra follow the same rule: their face half-edges are oriented, as each face forms a 2D subspace, and their faces have sidedness, as their 3D points generate force fields and a new location will be within one 3D zone or another.

Geographical objects are usually classified as 'fields' or 'objects', but here we see an alternative model where the primary entity is fields: all space is occupied by forces emanating from embedded generators. (If the field is global then the generator exists exterior to the map space – perhaps in a higher dimension – think of the centre of the earth when considering topographic elevation.)

2.3.6 *Boundaries: Two-sidedness and Voronoi cells*

Boundaries are formed where moving particles accumulate. There is a natural attraction to zones of heterogeneity, leaving an empty homogeneous void (a desert) in what we call a 'polygon'. This void is 'empty' only in the sense that no particles remain in its interior – it is 'full' in the sense that the force fields of its boundary particles (all of a like type) occupy it completely. Based on this idea, boundaries have 'sidedness', with a different type of particle on each side. We still have an overall space-filling field, but each portion will have a value associated with the particle type – a global field with discrete portions, as opposed to a continuous field with a single generator. This returns us to two-sided boundaries and adjacency between polygons, as well as the same properties for individual force fields, or Voronoi cells.

Thus Voronoi cells are the hybrid link between fields and objects, being both space filling and with identifiable generators (objects). They also have a property in common with coordinate systems: a hierarchical indexing system is readily developed, speeding up computer searching.

2.3.7 *Duality*

Duality is the expression of the idea that each element in one set of things is paired with a single element in another set of things. (For example, if we have one set of squiggly symbols called 'capital letters' and another set called 'small letters' then each element in each set has a paired element in the other set: given 'A' we know 'a', and given 'a' we can identify 'A' again.) This can be very important, as we can preserve a permanent link between each set of elements, and is especially useful with geometric data.

To be technical for a moment, Poincaré duality applies to n-dimensional oriented closed manifolds (for us 1D, 2D and 3D only). A k-dimensional element in this manifold has a matching (n-k) dimensional equivalent in its dual.

Thus in 2D planar graphs, 0D nodes have matching 2D regions in the dual – and 2D regions have 0D nodes in their dual. Also, 1D edges have matching 1D edges in their dual. (Remember 'dual' just means the 'other' network, so it can go either way.) Since this involves 2D entities, this applies to planar graphs, as these have regions, and not non-planar ones. Thus individual 2D regions can be 'hard-wired' to individual 0D nodes in the other graph, etc.

In 3D networks, 0D nodes have matching 3D volumes in the dual – and 3D volumes have matching 0D nodes in the dual. Since we are referring to 3D volumetric elements, occupying all the available space, we are referring to a 'cellular complex'. (Note that not all 3D graphs are cellular complexes.) Similarly, 1D edges have matching 2D faces in their dual, and 2D faces have matching 1D edges in their dual as well.

(It should be mentioned that 'planar graphs' or 'cellular complexes' imply a partition of the space in which they are embedded, so we are talking about a 'field' in the sense that all the map space is covered, but not a continuous one – it is made up of discrete portions, each with possibly different attributes, rather than a continuously smooth force field.)

These k and (n-k) dimensional elements are paired, so there may be fixed links between the elements in the pair. Note also that in both 2D and 3D, faces (and volumes) may be referred to by their dual nodes and edges, so all spatial entities may be saved as nodes plus edges (plus a reference to the dual). Finally, nodes do not have simple topological pointers, so all spatial relationships can be expressed as relationships between edges (together with pointers to the terminal nodes).

The Dual is the Context!

An interesting 2D example of alternate ways of modelling space is in terrain surface runoff modelling. In finite-difference modelling the surface is partitioned into 'buckets' to hold the water, and down-slope flow is estimated iteratively between adjacent buckets. The volume of flow is based on the common boundary between a pair of buckets and the slope between them – but if the buckets are adjacent the slope is clearly vertical! Duality is then invoked. Slope is estimated between the 'centres' of adjacent buckets, along the connecting edge: the dual of the bucket is the centre point, and the dual of the common boundary is the edge connecting the pair of centres. For a particular time step the flow calculation involves the water height in the buckets and the common boundary (primal) and the slope between the bucket centres (dual). Without at least an implicit duality no calculation is possible.

The partition of the surface into buckets may take various forms. If we have a sufficient set of observation points we may use these as bucket generators, using the concept of dominance: lacking any further information, the value (elevation) at any arbitrary location is probably that of the closest data point or, at the least, the average of all locations closer to that data point than to any other will probably be that of the data point. Thus a Voronoi diagram is appropriate. The connectivity between adjacent cells (and thus the slope estimates) is given by the dual Delaunay triangulation.

2.3.8   *Coordinates*

If the number or distribution of the data points is inadequate then pseudo-data points must be estimated by some other form of interpolation: the method should be able to take account of a wider selection of the data points. The location of these pseudo-data points may be random (leading to new Voronoi cells) or they may be 'regular' – based on some global coordinate system superimposed on the terrain. (The earlier slope calculations do not absolutely require a global coordinate system: a tape measure along the ground would suffice.)

We have now taken our terrain (a continuous field) and partitioned it into buckets (a 'discrete field' in our terminology). This is a start in our challenge of putting a model of space into the computer. However, the locations of our data points somehow need to be specified as well: a global coordinate system seems the obvious answer, although it raises several problems, as discussed earlier. Nevertheless, for data input and the output of results (graphics) this is a necessary evil. Luckily, as discussed earlier, functions such as CCW may be used, with coordinates as input and a (possibly guaranteed) decision as output: left, right, inside, outside, etc.

As seen earlier, these functions were derived from vectors, which are location-free: all we added were 'basis vectors' to give our frame of reference. This is as far as we can go to get a global-Cartesian-coordinate-free system.

In addition, we must make some decisions about the 'metric' we wish to use: this is usually the Euclidean metric, where the squared distance between two locations is the sum of the squared X-difference plus the sum of the squared Y-difference (plus the sum of the squared Z-difference in 3D). Other metrics may also be used, infrequently, such as the 'city-block metric' where the distance is defined as the sum of the X-difference plus the Y-difference.

If we then use this approach to specify the locations of our pseudo-data points we may end up with the familiar elevation grid, commonly used for terrain analysis – but it should be used with caution. We are now partitioning space by global coordinates, of a pre-defined spacing, and both the grid orientation and spacing may change with the merging of different data sets – it is always bad practice to re-interpolate based on pseudo-data that were already obtained by interpolation. The resulting grid is then used as before for our runoff analysis: the cells are the buckets and the dual edges are implicitly present connecting the grid cell centres.

We have now replaced 'adaptive' spatial partitioning based on the given data with spatial partitioning based on the superimposed coordinate-based grid. This has various advantages of simplicity, although it is moving ever further away from our true data. It is also convenient for spatial indexing and partitioning, e.g. k-D trees, quad trees, etc. (This may also be achieved with an adaptive partitioning, as will be discussed later.) This grid can not specify the identity of objects (an important issue in GIS) but only the attribute of some portion of the underlying object (if present) and awkward techniques are required to recover approximations of the underlying objects.

2.3.9   *The observer*

A final issue in the definition of a spatial model concerns the anticipated observer. How big must an object be to be relevant? (Blades of grass are irrelevant to us, but not to an ant.) Is the observer embedded in the map space? (If so, like a boat, it may bump into and interact with map objects, and if not, like a plane, it merely observes from overhead.) Are our entities capable of penetrating the boundary of an observed object? (Are boundary points close enough together that a point from an adjacent, but more distant, boundary is rejected as 'foreign', or alternatively is the boundary a solid line with a clear sidedness?)

2.3.10   *The computer*

We now have spatial models appropriate for the insertion of discrete fields into the computer, with the necessary introduction of a global coordinate system (basis vectors). Our other data type, discrete objects, may also be inserted with the use of global coordinates to specify point locations – either for simple points or else the boundary points of complex objects. However, these entities have no direct mechanism for the specification of spatial relationships – there is no dual. However, techniques exist, following Einstein, to expand the objects to fill the space available – the Voronoi diagram again. Then the dual, and basic spatial relationships (the context) are available for these new discrete fields. The initial question – what is here? – is replaced by 'what is closest to here'. The previously defined CCW test is used for this purpose: to locate the closest point or object and, for complex objects, to find if one is inside or outside.

2.4   SPATIAL DATA STRUCTURES

From our consideration of space we have concluded that we can model space in the real world like a force field, such as gravity, with objects (initially points) embedded in it. In our cosmic model points are generators of gravitational force, and other objects would be attracted to the nearest generator. Regions of low density would expand as remaining points migrate towards higher density areas – initially pancakes (faces) and, within them, towards edges at their intersections, and then towards nodes.

Objects would approach from one side or the other, but would not migrate through faces, making them impenetrable, giving a 'sidedness' to them. In the same way particles migrate within faces towards edges, and within edges towards nodes. 2D or 3D voids represent the near-absence of particles, but they are filled with the force fields of particles along their edges (2D) or faces (3D). These may be modelled as Voronoi cells.

Because of the sidedness and impenetrability of these faces or edges, data structures representing polygons or polyhedra (voids) must be able to represent individual sides – half-faces in 3D, half-edges in 2D. But we must be able to show where those half-faces or half-edges come from: we must be able to link them with the associated polyhedra or polygons on each side. Thus we have the dual relationship – the dual is the context.

A spatial data structure is an implementation of a spatial model within a computer. If we have 2D polygons, for example, following the 'cosmic' model we would need interior (half) boundary edges where pairs of polygons meet, together with matching boundary edges for those adjacent polygons – and dual edges expressing this context (polygon 1, polygon 2, boundary edge 1, boundary edge 2).

The well-known Quad-Edge structure (discussed later) takes advantage of the property of duality. Its pointer structure can be expressed as two hard-wired pairs of primal-plus-dual half edges – these are linked together by pairing up two half edges in the primal and two half-edges in the dual space to form a complete primal edge plus a complete dual edge.

The Dual Half-Edge structure, described later, follows the same approach in 3D: pairs of hard-wired primal-dual half-edge pairs are joined to form a complete primal edge plus a complete dual edge. The difference here is that the complete primal edge links up two adjacent geometric nodes whereas the complete dual edge links up two adjacent dual nodes – that is to say, two adjacent volumes. (In 2D the complete dual edge links up two adjacent areas in the same plane as the primal graph, while in 3D the dual edge is perpendicular to the face formed by the primal edges.)

The Voronoi spatial model fits with our Cosmic Model in that it can be used to describe what we think is going on: a set of objects whose force fields fill the available space, and whose zones of influence meet at boundaries – and whose relationships are described by the dual.

Putting it all together, we can develop a strategy for putting space into the computer. We create a pointer- or table-based implementation of a graph data structure to permit the storage, modification and query of the graph: computers are good at this. We admit a global coordinate system, via basis vectors, to permit a metric based determination of dominance, and hence a circumcircle-based determination of the Voronoi/Delaunay duality. We use the simple incremental algorithm (or one of the rather more efficient algorithms) to construct our VD/DT, in 2D or 3D. We then have a viable general method for handling space in a computer.

2.5   PAN GRAPHS AND QUAD-EDGES

However, before we can do this we need to translate these spatial entities and their spatial relationships into a structure we can code. We shall first look at spatial data structures in general, using the

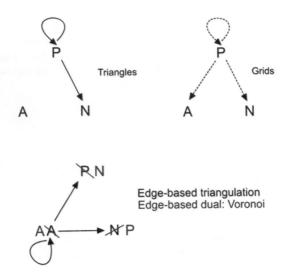

Figure 70. PAN graphs.

PAN (or VPAN) graph tool. Once we have decided on the properties that we want in our structure we can translate the results into a table-based or pointer-based implementation.

First we have to define the elements we want to preserve – often points (vertices), edges (arcs) and areas (polygons): each of these may be saved in the computer, along with their relationships with their neighbours. (In recent advances in full 3D models we may include volumes as well.) However, we probably don't want to store every connection to everything else: every adjacent node, arc and polygon to every node, and to every arc, and to every polygon – we will look at a simple tool to help us see which relationships we need for a particular application. We call this the PAN graph: it shows the chosen set of relationships between the three entities (P)olygons, (A)rcs and (N)odes.

The PAN graph is itself a graph – in fact a directed hypergraph. Not all of the entity types need be represented if you don't need them. Edges are unidirectional: a pointer from an arc to an adjacent node does not mean there must be a pointer from the node to the arc – this would be more difficult to handle, as while there are only two nodes adjacent to the arc there are multiple arcs adjacent to the node. PAN graph nodes may have edges to themselves: polygons can be adjacent to other polygons. The important thing is that one can navigate between the elements needed for the application. PAN graph edges may be real – they may be stored as pointers in a structure – or they may be implicit – they can be found out by other (usually geometric) means. Finally – which we will look at a little later – they can easily be changed to show the dual relationships in our spatial data structure. We will show some simple examples.

A simple introductory structure for a triangulation would have a file for triangles and another for nodes. As we know a triangle has three nodes, it is easy to put three node pointers in each record of a triangle file – but as the number of triangles around a node may vary, we will try to avoid putting pointers in the node file. However, we also know there are three triangles adjacent to a particular triangle, so we can also add three polygon (triangle) pointers to the records of the triangle file – as in the first diagram of Figure 70. With this we can 'walk' from triangle to triangle, as required for the point insertion algorithm described previously, and only referring to the node file when we need their X, Y coordinates. This is sufficient: but it does mean, when arriving at a new triangle, that we must check the edges to find out which one we came from.

A grid structure is perhaps the simplest of all: there are only squares (polygons) – arcs and nodes can be calculated when needed, and moving from one square to an adjacent one can be done by

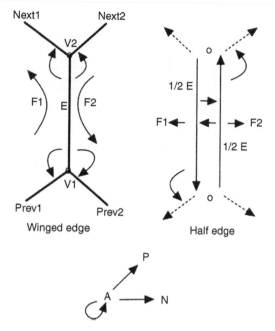

Figure 71.   Winged edge and half edge.

calculating its position from the number of rows and columns – an implicit relationship, shown by dashed lines, in Figure 70.

The third diagram in Figure 70 shows an alternative structure for the Delaunay triangulation (in black). Here the edge is the primary element, with pointers to the two adjacent nodes and the two adjacent triangles – as well as to 'adjacent' edges – by which we mean those edges clockwise or anticlockwise around either of the nodes at the ends of the edge – giving four more pointers. An interesting property is that this PAN graph also expresses the dual structure (in red). We saw earlier on that in 2D the dual graph is formed by exchanging nodes for areas, areas for nodes, and edges for the dual edges that intersect them: the equivalent dual structure is shown on the PAN graph simply by re-labelling its nodes. Thus the same structure can be used to represent either structure, allowing navigation from edge to edge around the triangulation, visiting triangles and nodes as required, or for navigating the dual diagram – or for any polygon map and its dual triangulation.

This avoids the problem described earlier: when, on entering a triangle, all three edges must be checked to find which edge was used to enter from. However, an equivalent problem arises: when a new edge is arrived at, it is unclear which 'side' it was entered from: was vertex 1 on the left and vertex 2 on the right, or the other way round? (See the 'winged edge' in Figure 71.) We still need to match the vertices of the old edge and the new edge.

An apparently definitive resolution of this conundrum is the 'half edge' and its variants (see Figure 71). Firstly each edge is 'sliced' down its middle, and pointers inserted to the 'other' half. Secondly an 'orientation' is given to the overall structure, e.g. the half edge points to the right, with a pointer to the associated vertex. Thirdly the pointer to the next half edge (anticlockwise) around the vertex is specified. (In some cases the clockwise pointer at the other end may be added as well.) This gives an 'edge algebra' structure, where navigation within the whole network may be specified by successive pointer calls, without any checks of triangle or edge orientation. The PAN graph for all of these is unchanged.

These structures may be implemented directly as pointers in memory, by tables with a row for each triangle, edge or half edge, or by tables in a relational DBMS, depending on the project requirements.

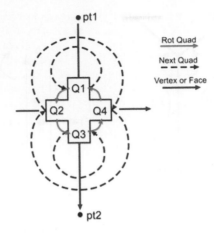

Figure 72.   The Quad Edge structure.

The Quad-Edge

In some cases an extra step may be added, based on the earlier observation that each edge of the dual graph intersects (or is matched by) a single edge in the primal graph. Thus the primal and dual graphs may be merged by providing pointers between these edges. The result is the Quad Edge structure, where each half edge – or 'Quad' – has a pointer to the next anticlockwise Quad, the primal pointing to the dual, then back to the primal etc. In the final structure therefore each Quad has a pointer to the next Quad around the final primal/dual edge (called 'Rot'), to the next Quad-Edge around the vertex (as in the half edge) and to the vertex itself (Figure 72). This allows navigation within the primal graph, within the dual graph, and from one to the other. Navigation from one complete edge to the next uses the 'Next' pointer. This is of value where both need to be accessed for some application – for example the flow model described previously – and also because it greatly simplifies the operations for the construction and editing of the graph. It requires more storage than the other options, as both primal and dual are maintained: the primal and the dual are updated at the same time. (Recently the same approach has been successfully applied to fully 3D models, with volume elements as well as faces, edges and vertices.)

The idea of MakeEdge is best illustrated by seeing it as a new element on a new shell (which would be an initially empty sphere – see Figure 73). The primal vertex loops would connect back to themselves, as there are no other edges connected to the vertex, and the dual vertex loops enclose the Voronoi regions associated with the primal vertices – hemispheres in this case.

The result of 'MakeEdge' is that a new primal/dual edge is created, ready for connection to the graph under construction. The pointers 'Rot', 'Next' and 'Vertex' are shown.

Maintenance of such a graph and its dual on a 2-manifold (such as a sphere or a torus, as well as a planar model with a 'back face') requires just two operations (or 'Operators'). The first is 'MakeEdge', which creates a single Quad Edge of four Quads, unconnected to anything else. The second is 'Splice', which either splits a vertex into two and merges two polygon loops into one, or else merges two vertices and their vertex loops while splitting the polygon loop into two (Figure 74). See Guibas and Stolfi (1985) for details.

The left hand side of Figure 74 shows part of a Quad-edge graph and the nearby quads to 'q'.

'q.rot' is the quad anticlockwise from q.
'q.rot.next' starts with 'q.rot' and points anticlockwise around the face (i.e. the dual vertex).
('q.rot.next.sym' would give the next quad clockwise around the vertex of 'q'.)
'q.next' gives the edge anticlockwise around the vertex of 'q'.
'q.sym' (i.e. 'q.rot.rot') gives the quad in the opposite direction from 'q'.
'q.sym.next' gives the next edge anticlockwise around 'q.sym'.

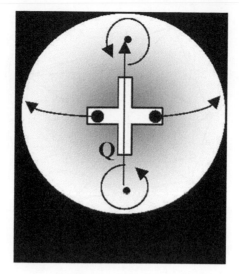

Figure 73.　A new Quad Edge.

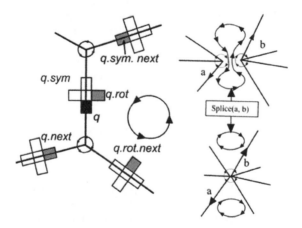

Figure 74.　The Splice operator.

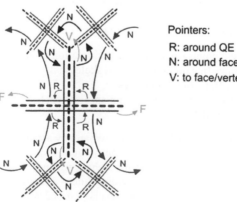

Pointers:
R: around QE
N: around face/vertex
V: to face/vertex

Figure 75.　A complete Quad Edge element.

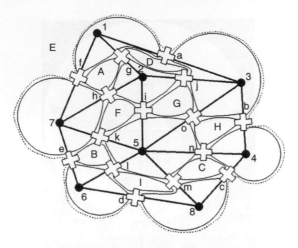

Figure 76.   A small Quad Edge triangulation.

The right hand side of Figure 74 shows the 'Splice' operator. The top part shows two separate vertices, with pointer loops around them. They are parts of a single face loop around the dual vertex. The bottom part shows the two vertex loops joined, and consequently the original face loop split into two. Repeating the 'Splice' operator reverts the structure to the original graph. The operator is achieved simply by switching loop pointers locally.

Figure 75 shows the complete set of pointers for a complete Quad-edge.

Figure 76 shows the 'Next' loops for a small graph – around each vertex and each face.

Navigation anywhere within the graph is thus possible by just following the pointers.

2.6   INTRODUCTION TO GEOGRAPHIC BOUNDARIES

What has this got to do with geography? It gives us a framework for stuffing our space into a computer.

A primal/dual pair is a single piece in a jigsaw: the primal part gives the colours/attributes of the piece; the dual part gives the wiggly boundary to be matched. The piece can be saved, and 'snapped' to neighbours. The primal is the object, the dual is the context.

Each half-boundary (and its portion of its void) is a jigsaw piece. It can be snapped to an adjacent half-boundary by using its dual. Half-boundaries have 'sidedness' – they have meaning/orientation only from their parent region.

We can now give a tentative classification of boundaries:

Class 0: individual independent points (including a consideration of points that move)
Class I: lines of points, or strings (including crusts and skeletons)
Class II: cluster boundaries, or 'fat' void boundaries
Class III: double-point boundaries (crust and skeleton applications)
Class IV: solid line segments – half-line pairs

Figure 77 emphasizes another aspect of boundaries: what they actually mean! Cartographically, boundaries separate polygons/areas/voids – but some polygons may exist though they have no well-defined boundaries. An example is forest stands, where we know that A and B have different tree types, but we can't actually draw a precise dividing line – a 'hard' boundary does not really exist. Another example is in flow analysis: the water levels of A and B are known at the observation points, and are assumed to be valid throughout the polygon, but the boundaries do not really exist – they serve to indicate flow between A and B, as indicated by the dual edge (often the slope – see

56

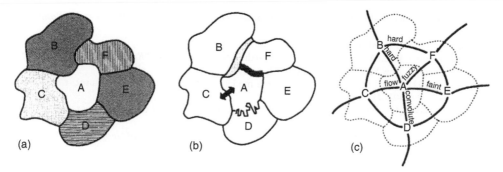

Figure 77.   Types of cartographic boundaries.

'Flow Modelling', below). In addition, boundary representations may be faint, convoluted, fuzzy etc. – again probably represented in the dual edge.

FURTHER READING

In the computer age, once we have spatial models we soon need data structures to allow them to be represented digitally. While the Newtonian (grid) model has been used since the beginning, a good survey, including hierarchical structures, is Samet (1984). These refer to field-type spatial models. Baumgart (1972, 1975) described a 'vector' data structure to express the relationships between the edges of a polyhedron.

A little later, considering Voronoi diagrams and Delaunay triangulations, Guibas and Stolfi (1985) – see below – combined the primal and dual graphs into a single structure, the quad-edge. This permits the navigation from the Voronoi to the Delaunay, and back again, and it is particularly elegant in that it requires only two primitive operations to maintain a valid structure.

Gold (1988) generalized the description of the various data structures using the PAN graph, in order to clarify the design choices required for a particular application. More recently Winter and Frank (2000) and Schneider (2009) catalogued the various data structures and their properties.

In addition, geographers have spent a great deal of effort in describing the nature of spatial structures, often from a computing perspective. Excellent examples are Goodchild (1992), Frank (1992), Laurini and Thompson (1999), Longley et al. (2001) and Worboys and Duckham (2004). Pullar (1994) looked in detail at the fundamental polygon overlay problem, and Albrecht (1996) attempted to classify the various GIS analysis operations. Galton (2004) examined the fundamental relationships between fields, objects and time.

Meanwhile, object-based models, initially for CAD systems, were being developed, for example by Baumgart (1972, 1975). These are edge-based structures.

REFERENCES

Albrecht, J.H. (1996) Universal GIS operations for environmental modelling. In: *Third International Conference/Workshop on Integrating GIS and Environmental Modelling, Santa Fe, New Mexico.*

Baumgart, B.G. (1972) *Winged Edge Polyhedron Representation.* Technical report, Stanford University Computer Science Department, Stanford Artificial Intelligence Report No. CS-320, 1972.

Baumgart, B.G. (1975) A polyhedron representation for computer vision. *IFIPS Conference Proceedings*, 44, 589–596.

Clarke, S. (1717) A Collection of Papers, which passed between the late Learned Mr. Leibniz, and Dr. Clarke, In the Years 1715 and 1716, by Samuel Clarke D.D. London, James Knapton.

Coles, P. (1990) Understanding recent observations of the large-scale structure of the universe. *Nature*, 346, 446–447.

Descartes, R. (1728). *Le monde, ou Traité de la lumière.*

Einstein, A. (June 9, 1952) As quoted in the Preface to the fifteenth edition of *"Relativity: The Special and the General Theory"*. New York, NY, Crown Publishers, 1961. p. vi.

Frank, A.U. (1992) Spatial concepts, geometric data models, and geometric data structures. *Computers & Geosciences*, 18, 409–417.

Galton, A. (2004) Fields and objects in space, time, and space-time. *Spatial Cognition and Computation*, 4, 39–68.

Gold, C.M. (1988) PAN graphs: An aid to G.I.S. analysis. *International Journal of Geographical Information Systems*, 2, 29–42.

Goodchild, M.F. (1992) Geographical information science. *International Journal of Geographical Information Systems*, 6, 31–46.

Ikeuchi, S. & Turner, E.L. (1991) Quasi-periodic structures in the large scale galaxy distribution and three-dimensional Voronoi tessellation. *Monthly Notices of the Royal Astronomical Society*, 250, SlS.522.

Icke, V. & van de Weijgaert, R. (1991) The galaxy distribution as a Voronoi foam. *Quarterly Journal of the Royal Astronomical Society*, 32, 85–112.

Laurini, R. & Thompson, D. (1999) *Fundamentals of Spatial Information Systems*. London, Academic Press.

Longley, P.A., Goodchild, M.F., Maguire, D.J. & Rhind, D.W. (2001) *Geographic Information Systems and Science*. London, Wiley.

Pullar, D. (1994) A tractable approach to map overlay. PhD Dissertation. University of Maine at Orono.

Samet, H. (1984) The Quadtree and related hierarchical data structures. *ACM Computing Surveys*, 16, 187–260.

Schneider, M. (2009) Spatial data types. In: Liu, L. & Özsu, M.T. (eds.) *Encyclopedia of Database Systems*. Berlin, Springer-Verlag.

van de Weijgaert, R. (1994) Fragmenting the universe III. The construction and statistics of 3-D Voronoi tessellations. *Astronomy and Astrophysics*, 283, 361–406.

van de Weijgaert, R. (2007) Voronoi tessellations and the cosmic web: Spatial patterns and clustering across the universe. In: *Proceedings, ISVD 2007*. pp. 230–239.

Winter, S. & Frank, A.U. (2000) Topology in raster and vector representation. *GeoInformatica*, 4, 35–65.

Worboys, M. & Duckham, M. (2004) *GIS: A Computing Perspective*. Boca Raton, FL, CRC Press.

Yoshioka, S. & Ikeuchi, S. (1989) The large scale structure of the universe and the division of space. *The Astrophysical Journal*, 341, 16–25.

Zaninetti, L. (1991) Dynamical Voronoi tessellation III. The distribution of galaxies. *Astronomy and Astrophysics*, 246, 291–300.

Chapter 3

Points

3.1 SINGLE-CELL ANALYSIS: LABELLED AND UNLABELLED

We have seen how dominance, the assignment of spatial locations to the closest generator, leads to a fundamental spatial partition – the Voronoi diagram. In some cases only the generator location is known, and in others some attributes, such as colour, may be assigned.

In forest mapping, the Voronoi diagram clearly gives a good estimate of the context of an individual tree in a forest: its constraining neighbours are found directly if tree crowns may be detected from aerial imagery. The locations of the tree crowns may then be used as generators, and the area of each associated Voronoi cell gives a reasonable first estimate of the tree's spread, in competition with its neighbours, and tables may be constructed to compare this with the tree height and wood volume. (This applies where the trees are in close competition, and does not apply at the edge of the forest stand.) Summing these volume estimates for a particular area may be used to estimate harvestable timber. In some cases the tree type may also be estimated, and the resulting volume totals used for harvest planning.

The same approach may be used for other types of natural resource assessments. Coal thickness for example, or permeable aquifer thickness, is often assessed by shallow drilling – and the total thickness observed at each location. (This may often involve several thin strata, and often only the total thickness is needed – this is difficult by modelling the geological 'tops' of each stratum. Experience suggests that stable regional estimates may be made by assuming the local thickness at any location is best estimated by the closest real observation – the Voronoi cell generator. Summing the thickness times the cell area often gives satisfactory results for minimum effort. The same holds true for total rainfall estimation based on rain gauges – as originally suggested by Mr. Thiessen.

A related, and common, problem is that of point density mapping. This is usually attempted by counting circles – counting how many points fall within a given circle, and then moving it and trying again: a very small circle, or a very large one, would give implausible results. Noting that point density (points per unit area) is the reciprocal of the Voronoi cell area (area per unit point) greatly simplifies this process. In this case the local density around each generator is 1/(cell area). Mapping this directly will often give a highly variable density surface, and some traditional smoothing function is often appropriate. Nevertheless, the whole problem of using counting circles has been eliminated.

A significant application of this approach may be found in the work of van de Weijgaert described earlier. A major catalogue of a significant portion of the cosmos was performed by the 2dF Galaxy Redshift Survey, completed in 2002, extending over a huge number of galaxies, and light-years, in 3D (Figure 78). In order to speculate on the cosmic structure they needed to produce a density map rather than just galaxies/points, and so they developed the 'Delaunay Tessellation Field Estimator' procedure. Essentially this consisted of producing a Delaunay tetrahedralization of the galaxies, calculating the total volume of those tetrahedra adjacent to each one – and then using the reciprocal as the density measure. (They could just as well have used the volume of the associated Voronoi cell, as this would have produced very similar results.)

The same approach models a large number of natural processes: examples are turtle shells, crocodile skins, mud and basalt cracks and many more.

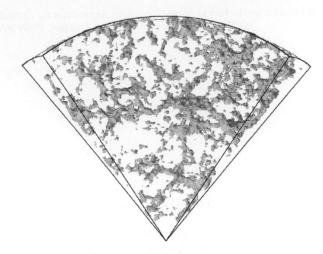

Figure 78. 2dF Galaxy Redshift Survey.

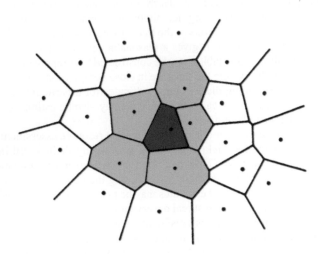

Figure 79. Voronoi neighbours.

3.1.1 *Context*

It can be argued that one of the main problems of spatial analysis – the knowledge of the neighbourhood of an object – can be effectively managed by using the Voronoi diagram. We call this the 'Context' of the object (Figure 79).

Thus the context – the neighbourhood – of the point in the red cell in Figure 79 is the set of (green) Voronoi neighbours: any minor perturbation of the generator of the red cell, or even its deletion, will only affect the green cells. (Conversely, if P was not initially there, its insertion will only affect the green cells.) This structure resolves the 'post-box' problem: to find the nearest point (post box) to any spatial location, as each Voronoi cell contains all the spatial locations closer to its generating point than to any other.

3.1.2 *Nearest-point interpolation*

If the Voronoi cells are considered, another elementary interpolation method is seen: all locations inside the cell – that is, closest to the central data point – have the height of that point. (Raster

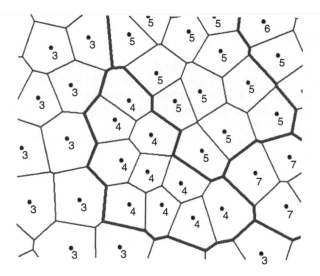

Figure 80.   Labelled skeleton.

cells, for example, are considered to have the same elevation throughout, equivalent to a central elevation point.) This will give a rather strange blocky model for elevation, but it is frequently used in remote sensing, where each pixel must be assigned a colour intensity, but averaging the intensities of several neighbours may blur the values of the objects being identified. Occasionally a pixel should take the value of its nearest data point rather than averaging a set of neighbours: keeping to a single (nearest) neighbour helps preserve the object colour for identification purposes. The VD directly provides this function.

3.1.3   *Labelled skeletons*

In the previous cases generator attributes are often based on the continuous number scale, and are not appropriate for classification of generators and cells. In other cases attributes are 'nominal', describing for example different rock types. Here we may often assume that the dominance boundaries between generators with the same label are less important than between rock types, so they are suppressed, giving us 'labelled skeletons' – topologically connected boundaries around each cluster of similar labels. Preliminary geological mapping is a good example: in many areas the geologist has to hunt to find identifiable outcrops that are not covered with overburden, and label them as to the rock type. As nothing is known in the covered regions the best guess is the closest observed outcrop – hence the Voronoi cell can form the basis for the initial map-colouring exercise (Figure 80). This is a simple form of interpolation, although the variable being plotted is not continuous. These maps may then be modified by the expert to estimate formation boundaries based on additional geological knowledge. We refer to this as the 'labelled skeleton' to distinguish it from the 'geometric skeleton' described below, where the skeleton boundaries are derived without manual point labelling.

Another application is in maritime boundary delineation, where the 'half-way' rule is used to separate pairs of jurisdictions – see Boas (1951) and Ricketts (1986).

It should be noted here that, as a result of the interlinked Voronoi/Delaunay relationships, the topological connectivity of the cell boundaries is guaranteed to be complete – and so also will be the labelled skeleton around a contiguous set of Voronoi cells. Referring back to the description of the Incremental Algorithm, it can be noted that the only operations that change the connectivity of the data structure are 'Insert' and 'Swap' – and these are done by a guaranteed set of operations based on the Quad-Edge 'Splice'. This is an extremely important concept that we will use in other applications later on.

FURTHER READING

Alfred Thiessen was a meteorologist who developed the method for the analysis of spatially distributed data. The term 'Thiessen polygons' is often used in geography and meteorology. (Score one for meteorology.) The key idea was to assign the rain gauge results to the whole proximal area: while not precise at any point location, it provides the best known estimate for the sum of all locations in the cell.

Within metallurgy, Voronoi diagrams are often referred to as Johnson-Mehl diagrams, after Johnson and Mehl (1939).

G.S. Brown (1965) used the same approach for point density analysis in forestry. Stems/acre is the reciprocal of acres/stem – i.e. the area of each cell. Pelz (1978) used the same approach for estimating individual tree growth. (Score one for forestry.)

A variety of other applications preceded or soon followed in the natural sciences, e.g. Smalley (1966) for basalt flows, Mead (1967) for inter-plant competition, Saito (1982) for epidermal cells, Arnold and Milne (1984) for soil surveys, and Boas (1951) and Ricketts (1986) for delineating maritime boundaries.

3.2 TRIANGULATION – TIN MODELLING

Simple interpolation

While plausible detailed interpolation requires the evaluation of several neighbours, the simple algorithm previously described can be surprisingly useful: first of all construct the Delaunay triangulation; then select the locations to be interpolated (perhaps on a grid); then walk to each one in turn to find the enclosing triangle, while using the final CCW results to estimate the elevation of that location within the enclosing planar triangle. Alternatively, if a simple contour map is what is desired, just take each triangle in turn and calculate its intersection with the various contour levels. While not elegant, it gives a rapid surface model. The Voronoi cells could also be examined, and might be used as a simple runoff model.

Figure 81 is a simple point data set where the values equal the elevation.

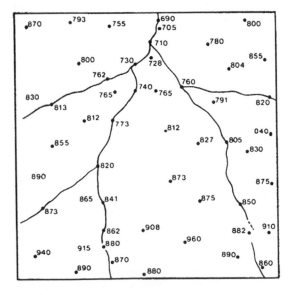

Figure 81.   A simple terrain dataset.

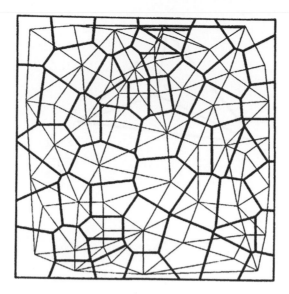

Figure 82.   VD/DT of terrain dataset.

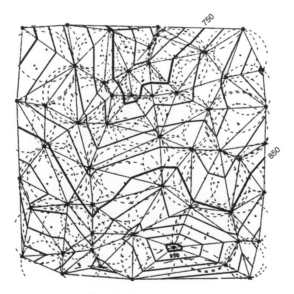

Figure 83.   Linear interpolation of terrain dataset.

The point set can be triangulated with the Incremental Algorithm, and both the Delaunay triangles and the Voronoi cells drawn (Figure 82).

It is easy to imagine that each triangle is a flat plate, giving a simple terrain model – the commonly used TIN model – and the lines drawn at the intersection of the contour level and the triangular plate. (In addition Figure 83 shows that all the circumcircles around the Delaunay triangles are empty, and do not contain any interior data points.) This gives a basic terrain model.

If we return to TIN terrain modelling, we need to be able to estimate the height at any intermediate location. As with the Incremental Insertion Algorithm we need to be able to 'Walk' through the triangulation to find the triangle enclosing our desired location. This automatically gives us the

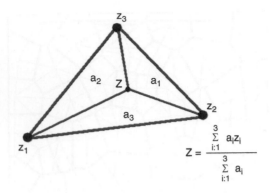

$$Z = \frac{\sum\limits_{i:1}^{3} a_i z_i}{\sum\limits_{i:1}^{3} a_i}$$

Figure 84.   Planar interpolation within a triangle.

Figure 85.   Contour data points.

three positive CCW areas of our location point with respect to the triangle: these may be used directly to perform planar interpolation within the triangle (Figure 84).

For much of our discussion of terrain interpolation we will use contour data as our example. This is for two reasons: firstly it is still a very common source of data for small-scale projects, and secondly the resulting surface is easy to visualise as we describe the strengths and weaknesses of various interpolation techniques. Figure 85 is a set of contour points from an imaginary terrain surface.

Figure 86 is the triangulated result. If we interpolate on the basis of flat TIN triangles we get Figure 87.

In many places this result is plausible, but in a few locations – where triangles have all three vertices on the same contour and thus have the same elevation – the results are clearly bad. We will examine these 'Flat Triangles' in a later section.

3.3   PASSIVE INTERPOLATION: THE PROBLEM

We now look at the more usual question: how to estimate the height at any location on a terrain model based on examining several nearby data points. (We will ignore more global techniques, such as 'trend surface analysis', where some mathematical function is fitted to the whole map area.)

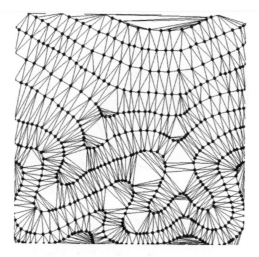

Figure 86. Triangulation of contours.

Figure 87. Simple TIN interpolation.

The largest class of these local techniques is the 'weighted average' method. This has three steps for estimating the elevation at some chosen location P: a) selecting a set of neighbouring data points to use for the average; b) specifying the elevation function being averaged (often being the elevation alone, but possibly involving the slope at the data point); and 3) defining the weighting function, so more important points have more influence.

We have already seen two simple weighted average techniques: the trivial 'nearest neighbour' method, where the closest point to P gets all the votes, and the triangulation (TIN) method, where the three vertices of the triangle enclosing P are weighted using the CCW values – see Figure 84. The elevation of P is estimated as the sum of the three vertex elevations times their 'weights' or areas – then divided by the sum of the areas alone (which equals the area of the enclosing triangle). As this is a linear combination of the vertex Z values, the result is a planar interpolation function inside (and outside) the triangle – what you would expect for a TIN model. We can also see that as P approaches Z_1 its associated area a_1 approaches 100% of the big triangle area, so our interpolant passes through the data points and matches them precisely. In addition, as P approaches the opposite

65

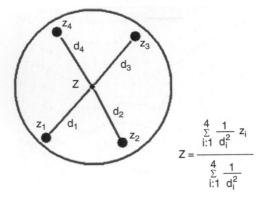

$$Z = \dfrac{\displaystyle\sum_{i:1}^{4} \dfrac{1}{d_i^2} z_i}{\displaystyle\sum_{i:1}^{4} \dfrac{1}{d_i^2}}$$

Figure 88.   Inverse distance weighting and counting circle.

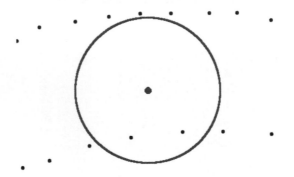

Figure 89.   Counting circle and contour data.

edge, a_1 approaches zero: the same thing would occur for the adjacent triangle, so values along the edge would not be influenced by Z_1 – and the surface would be continuous across triangle boundaries. Slopes, however, would not be continuous across triangle boundaries and vertices, as even close to the edge the surface within our triangle would be influenced by Z_1, whereas the adjacent triangle surface would be influenced by its far Z value.

However, for smoother or better looking surfaces more points are needed, as well as a more flexible weighting function. A traditional, but problematic, method is 'inverse distance weighting', or the 'gravity model', based on the supposed similarity between the influence of a single data point on its neighbourhood and the force of gravity (which decays as $1/d^2$ where d is the distance between the data point and the interpolation location, although other exponents have been used). There is no obvious set of data points to use, so one of a variety of forms of 'counting circle' is used, as Figure 88.

When the data distribution is highly anisotropic there is considerable difficulty in finding a valid counting circle radius: for example where the input data comes from contours, as below, or from flight lines or ships' traverses: the data density along the track is much greater than the distance between tracks, making the selection of an appropriate counting circle difficult, and producing a poor distribution of data points around the location of interest. This can seriously bias the estimate towards one side of the data set (Figure 89).

Various attempts have been made to ameliorate this problem, for example selecting one point from each quadrant, or octant – see Figure 90. Case a) selects the nearest 6; case b selects all within a specified counting circle; case c) selects all within a square region; case d) selects the nearest in each octant. Many interpolation programs offer some of these choices, but the underlying problem remains: how to guarantee that the selection of neighbouring data points matches the weighting function, so that there are no discontinuities when the counting circle, or octant search, etc., adds

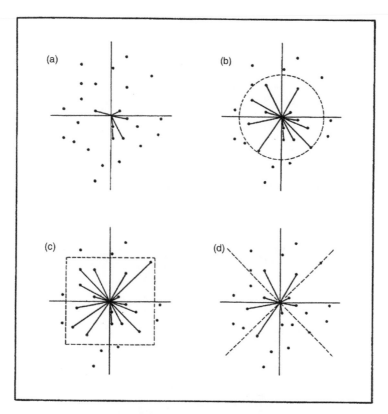

Figure 90. Neighbouring point selection strategies.

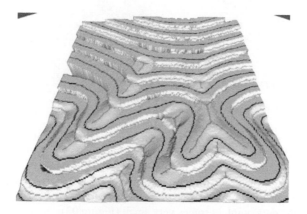

Figure 91. Interpolation with a small counting circle.

or removes a point as the query location is perturbed – the weighting function should reach zero by the time this happens. Similarly, when the query location precisely matches a data point, its weighting function must equal 100% (and the others 0%) or the surface will not pass precisely through the data point: here we assume the data is precise, with no error component.

Some simple examples will show the problem. For convenience we will use the contour data set shown previously, as it is easy to visualize, together with the gravity weighting model and counting circles of varying radii. In Figure 91 the radius is less than half the contour spacing: the result is a step-like function as the circle will only capture points from a single elevation – and in one case

Figure 92.   Interpolation with a large counting circle.

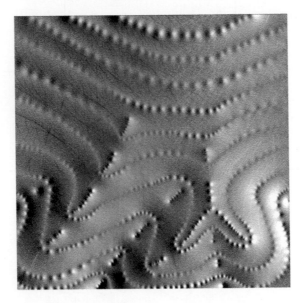

Figure 93.   Large counting circle: illuminated view.

on the left the wide contour spacing means the circle catches no points at all. (Here we have added the 'skeleton' (in red) between contour pairs – this will be discussed in a later section.)

For larger circles (Figure 92), greater than a quarter of the map size) the surface is smoother as more points are captured – maybe a large portion of the total for each query point – but pronounced bumps can be seen at each data point, since $1/d^2$ becomes very large when d, the distance between the query point and the data point, becomes very small (Figure 93).

In conclusion, because the selection criterion is metric, it is unable to adjust for varying point densities, while making the radius very large includes far too many distant – and irrelevant – points, greatly increasing the cost. Surface discontinuities occur when the circle is locally too small, and slope discontinuities occur when the centre is very close to a data point – because the weighting function increases exponentially.

At first sight it would seem that a precise elevation estimate is the primary purpose of interpolation, but in many applications the slope and aspect of the surface are more important – for questions of vegetation, insolation and trafficability for example. We therefore also need to look at methods that directly take account of the local data point distribution. Based on our previous discussions, the Voronoi/Delaunay methods seem appropriate.

FURTHER READING

Perhaps the first attempt at computer-assisted contour maps was by Bengtsson and Nordbeck (1964), where a coarse grid was superimposed on the map, weighted average values were estimated at the grid nodes and contour lines then threaded through the cells.

In the 1970s various workers in multiple disciplines were developing triangulation techniques for 2D monotonic surface interpolation. Gold (1979) produced one snapshot in the thick of the battle. There were two issues: how to define (and implement in a data structure and algorithm) a 'good' triangulation, and how to interpolate within each triangle. We now know (Shamos 1975) that only the Delaunay triangulation has a unique solution for any input order. The approach of Lawson (1977) of maximizing the minimum angle in the quadrilateral formed by a triangle pair was correct, using a Delaunay property, as was Sibson (1978, 1980) and Green and Sibson (1978). Gold et al. (1977) was in error by maximizing the minimum triangle height, but was useful for specifying the basic triangle data structure, the incremental insertion algorithm and its efficiency, and for defining the 'walk' algorithm to find the enclosing triangle – based on barycentric coordinates – as well as obtaining simple planar interpolation. He also defined a general triangle ordering based on an exterior viewpoint – see also Gold (1987).

Various triangle interior interpolation techniques were attempted, such as Akima (1978), but none were entirely satisfactory. Peucker et al. (1978) summarized their work on Triangulated Irregular Networks (TINs). Philip and Watson (1982) described an improved triangulation approach. Lee (1991) compared existing TIN algorithms. Gold (1989) mentioned previously – described an 'area-stealing' approach by inserting the query point into the triangulation, but this was preceded by 'natural neighbour interpolation' described in Sibson (1982). This is still the method of choice, perhaps with the addition of slope estimates at the data points. Li et al. (2005) gave a recent survey of terrain modelling.

Various historical surveys of interest are Peucker and Chrisman (1975), Rhind (1975) and Lam (1983).

3.4   THE DYNAMIC ALGORITHM: POINT DELETION

We have already discussed the Delaunay-based TIN model, which is simple to generate and provides a surface that passes through the data points – at the cost of frequent slope discontinuities at triangle edges and data points. We will now look at a Voronoi-based model.

Having the Delaunay tessellation of the data points means we also have its dual available as well, which gives us a good definition of neighbours to any point in the set. However, this set does not include the query location where we wish to make an elevation estimate: it must be added (and then removed) for each location of interest – at each point on a grid, for example. (The insertion of the query point provides a consistent definition of the neighbourhood – context – of the query point: this will be found useful in subsequent algorithms.) Incremental point insertion has already been discussed, but not deletion: this algorithm will be described now.

In the insertion process the triangle containing the new point is found, and split into three, replacing one triangle with three. However, the neighbouring triangles may contain a point within the circumcircles of each of the new triangles: these must be tested and, if that is the case, the common diagonal switched to conform to the Delaunay criterion. This leaves the new outer edges untested, and these are maintained on a stack and tested in their turn until all external edges are valid. Deletion is the reverse process: adjacent triples of 'potential' triangles surrounding the point P to be deleted are tested, to find one whose circumcircle is empty (except for P): the common diagonal is then switched back, creating a new exterior triangle that will remain when P has finally been deleted (Figure 94). Clearly any such triangle may not have any of P's neighbours inside its circumcircle. The process is then repeated until P has only three adjacent triangles, at which stage

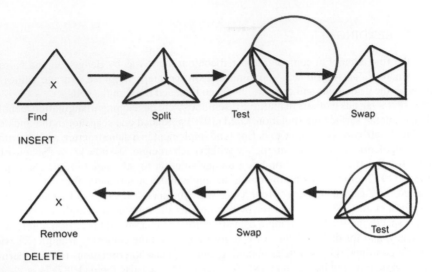

Find   Split   Test   Swap

INSERT

Remove   Swap   Test

DELETE

Figure 94.   DT point deletion algorithm.

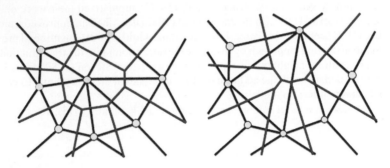

Figure 95.   VD/DT point deletion.

these may be removed – exactly the reverse of the insertion process. (Care must be taken in case four or more neighbours of P are co-circular: usually shrinking the circumcircle by a very small value 'epsilon' removes this problem.)

This allows any point to be inserted or deleted at will, giving a 'dynamic' data structure (Figure 95). (This term does not refer to point movement, but to any data structure that may be updated locally by the addition or removal of elements.) Figure 96 gives another example: note the circumcircles on the left – if they contain the new point location then the diagonal must be switched.

Complete algorithm:

Point Deletion
Let 'P' be the point to be deleted and V1, V2, V3 be three anticlockwise neighbours forming a potential triangle.

a) Skip if |V1 V2 V3| <0 (The triangle is re-entrant)
b) Skip if |V1 V3 P| <0 (P is inside the triangle V1 V2 V3)
c) Skip if any remaining neighbour of P is inside the circumcircle of V1 V2 V3
   i.e. H (Vi Vj Vk P) <0 (The circle is not empty)
d) Replace edge V1 V3 with edge P V2
e) Reassign V1, V2, V3 and continue until only three neighbours remain.
f) Remove P.

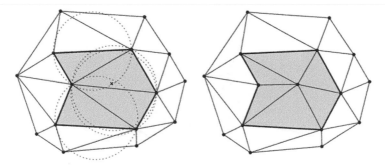

Figure 96.   DT point insertion and deletion.

N.B. To avoid having >3 neighbours on the circumcircle, shrink the circle radius by a very small epsilon.

FURTHER READING

The simple static Voronoi diagram is quite sufficient for many applications. The theoretically most efficient are 'batch' processes, where all points are entered at the same time, such as the sweep line algorithm of Fortune (1987), mentioned previously. Incremental algorithms are typically somewhat slower, but may be more stable, and allow the addition of subsequent points – as in the skeleton algorithm described later. However, many operations are facilitated if the algorithm is 'dynamic' – that is, points may be both inserted and deleted as required. Bajaj and Bouma (1990) provided an early outline. Heller (1990) provided a detailed deletion algorithm that was shown to be incomplete by Devillers (1999, 2002) – who provided a valid algorithm based on power diagrams. Mostafavi et al. (2003) provided a simple and effective algorithm based on the circumcircle, which is described further in this book.

3.5   INVASIVE METHODS: SIBSON INTERPOLATION

Let us look again at the concept of 'neighbour' in a Voronoi world (Figure 79).

A particularly important application of the Dynamic Algorithm is in Sibson interpolation (also called 'area-stealing' and 'natural neighbour'). The query point P is fully inserted into the VD/DT of the dataset, and then removed again, allowing us to calculate the areas stolen by P from each of the Voronoi neighbours. The weights consist of the areas of the neighbouring points' Voronoi cells that were taken over by the query point P. As before, the elevation (Z-value) is estimated as a weighted average (Figure 97).

Here we are using the 'boat' model, where the observer interacts with the existing map and generates a new cell – unlike the 'plane' observer, who only observes from above.

This is really an application of 'higher order' Voronoi diagrams, where we partition space into tiles with the same first-nearest and second-nearest generating points. This is often useful for questions of the form: "Which is the closest hospital to a location P, and which would be the next closest if it shut down?" In Figure 97 a query point P near hospital Z would fall within Z's Voronoi cell, and thus Z would be closest. However, if Z was removed then the original Voronoi cells would take over, and P would fall in the Voronoi cell of Z5 – and thus the patient P would be sent to hospital Z5.

Incidentally, this idea works for other 'field' models of space, where we have a tessellation covering our map. Figure 98 shows a US map with 'Kansas' shaded in. Frequently US social scientists have information aggregated by state, which they wish to use for some analytic operation – but what to do if some information is missing or lost? As a possible response, imagine that each state

71

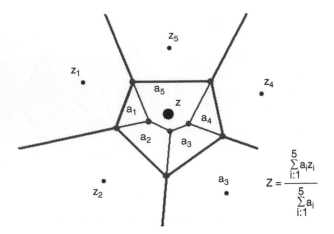

$$Z = \frac{\sum\limits_{i:1}^{5} a_i z_i}{\sum\limits_{i:1}^{5} a_i}$$

Figure 97.   Sibson interpolation.

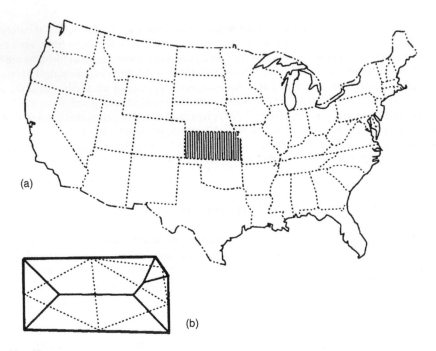

(a)

(b)

Figure 98.   Kansas.

is a Voronoi generator of the shape and size of that state: Voronoi regions would thus have a zero extent outside its border – the boundaries would follow the 'cracks' in the map, as in diagram 'a'.

Following our previous insertion-deletion approach, we would remove Kansas, leaving a Kansas-shaped hole. The neighbouring states would march in to occupy it, forming the Voronoi regions shown. Replacing Kansas would 'steal' these regions, giving the stolen areas used in our previous interpolation (diagram 'b') – and thus allowing a weighted-average interpolation of the missing Kansas data: perhaps population. While crude, this would allow some plausible replacement of the missing data – better, for example, than weighting by boundary length, which could be greatly distorted if a major boundary was formed by the Mississippi River!

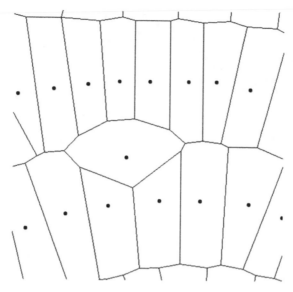

Figure 99.   Voronoi cells for contour data.

Returning to point-based interpolation of elevation data, the advantages of the Sibson approach come from the integration of the weighting function and the selection of the set of neighbours: a data point is a neighbour if it has a common Voronoi boundary with P, which can only happen if some 'area' has been stolen. Thus no surface discontinuities can occur as the set of neighbours changes when the sampling point P moves: at the moment when a neighbour is added or removed the area stolen is precisely zero. In addition, as P approaches a data point, the area it steals is taken more and more from that data point exclusively, so that the weighted average interpolated surface precisely matches all data points – although the surface slope varies with the direction from which the point is approached. It can be shown that the surface produced by Sibson interpolation is smooth (C^1 continuous) everywhere except at the data points themselves.

Figure 99 illustrates why: an interpolated point can never have weighted neighbours that only come from one contour: the central point steals cell area from both strings of contour elevation points (unless we have a sharp bulge, which will be discussed later). The same is true for other anisotropically distributed data sets, such as airborne surveys, ships' traverses or, in vertical sections, oceanographic soundings and temperature profiles. (In addition, this also works in full 3D: the stolen volumes from, say, a series of ocean temperature profiles would allow volume/area-stealing interpolation at any location (perhaps on a 3D grid) – the results would be 'plausible' without invoking any counting spheres, semi-variograms or other devices – such as in 'Kriging', below.)

An interesting case from a few years ago illustrates the value of the approach. Figure 100 shows the bathymetric observations around an artificial island in the Beaufort Sea, obtained from a small boat. (In fact, the upper right shows the demise of the boat – it collided with the caisson and sank! Luckily the data – and the crew – were rescued, and one arrived at my house owning only a life jacket and a data tape – demanding that I plot it immediately! Caulfield et al. (1984).

Most interpolation methods, based on searching for a set of neighbours within some counting circle, would have been unable to produce a map – but the Voronoi approach, being adaptive with respect to the data distribution, was able to do so, as shown in Figure 101. Other data types – ships' traverses, airborne surveys, on-foot data collection – have similar requirements.

We will look at Sibson interpolation again a little later, with the example of contour-line data, when considering our first type of boundary – strings of points.

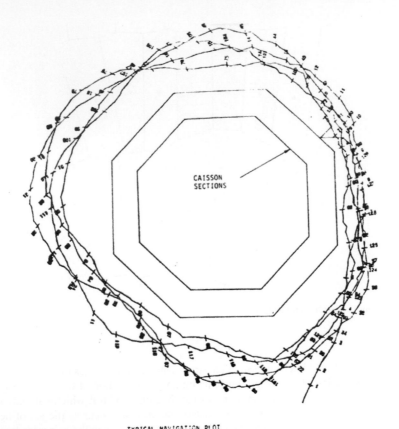

TYPICAL NAVIGATION PLOT

Figure 100.   Beaufort Sea caisson survey.

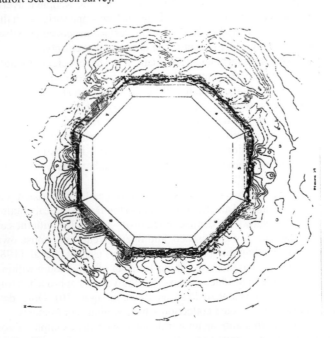

Figure 101.   Beaufort Sea contour map.

3.6 POSTSCRIPT: ACCURACY – THE KRIGING APPROACH

Apart from the rather ad-hoc jackknife method, when the original elevation at a data point is compared with an estimate at the same location made when the original data point is removed, a second approach which is particularly popular with geologists attempts to tackle the underlying conundrum head-on: there is often a significant error in the observations themselves (particularly for attribute data such as chemical content, porosity etc.) – which implies some statistical description. But this can only be obtained by repeated sampling – which is not possible for a single-valued surface. The Kriging idea starts by postulating the concept of 'regionalized variables' – where the variation at a single location, although unobserved, may be estimated from nearby locations: this assumes that localized variables behave in a similar fashion nearby, and the underlying variability at one data point may be estimated from its neighbours.

Kriging is also a weighted-average method, but here the weights are based on a previous analysis of the data set: how closely correlated are the data points? Of course, it would be expected that points would be most highly correlated when they are close together, and less so with greater distance: the first stage of Kriging is to make an estimate of this correlation function, expressed as a 'variogram', and then use it to estimate the weights for a particular point P and its neighbourhood set S.

There are a lot of assumptions involved here. They are based on the idea of 'regionalized variables' (RV): that is, a variable intermediate between a random variable (of the kind used in conventional statistics) and a deterministic variable (i.e. one without any random component, such as the methods discussed above). RVs are assumed to vary in a continuous manner from one location to the next – thus points that are near each other have a certain degree of spatial correlation, but widely separated points are statistically independent. RV theory assumes that the spatial variation in the data is homogeneous across the surface – that is, the same pattern of variation (correlation) would be homogeneous at all locations on the surface. It is easy to imagine cases where this is not true, so a statistical model must be built and tested first.

Potentially difficult cases include study areas with significantly differing terrain types – for example, in Canada and the US, for a map area with the Rocky Mountains to the west and the Great Plains to the east, the expected difference in elevation over 1km on the ground would be much greater in the mountains. Another situation would be a long ridge: variation along the ridge would be much less than across it. Finally, there is the question of 'stationarity': it is assumed in RV that the expected average value would be about the same wherever it is examined on the map: consistently sloping surfaces are not stationary, and the underlying 'trend' must be subtracted before Kriging methods may be applied. (This is often called Universal Kriging.)

Could one calculate how the correlation between data points varies with distance? If we ignore distance then a simple correlation coefficient can be calculated for the elevation values for the whole map. Spatial statistics uses a similar concept based on the 'intrinsic hypothesis' of Olea (1984). The intrinsic hypothesis states that outcomes from different locations can be considered as elements from the same population: this gives us a 'regionalised variable", which is thus considered to exist at any location on the map. If x and x + h represent two locations on the map separated

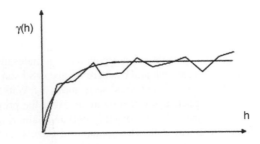

Figure 102. Experimental and model variograms.

by a distance h, and Zx and Zx + h are the random variables at these locations, then the difference [Zx − Zx + h] is another random variable. The intrinsic hypothesis is said to be satisfied if the first two moments of this difference are independent of the location and only depend on the distance h.

The first moment (like the mean in ordinary statistics) states that the expected value of the differences is the mean for the specific separation h: $E[Zx − Zx + h] = \mu h$ – where E means the expected value.

The second moment (like the variance in ordinary statistics) states that the variance of the difference [Zx − Zx + h] is the difference between the variances Zx and Zx + h divided by 2. (Note that, as in ordinary statistics, the mean must be subtracted from the variables first.) Thus if Z has a constant mean μ the theoretical variance $2\gamma x$, $x + h = E[(Zx − Zx + h)^2]$ – see Wackernagel (2003), Cressie (1993).

However, this could only be used directly if every location on the map had an observation and every pair of points, with all possible separations, was used: not very practical. To use Kriging in practice we must first group pairs of elevations in our real data set with similar values of h and then plot the Experimental Variogram for each group. In Figure 102 the various separations h for a 'training set' of data points have been combined into several groups and the variance has been calculated for each interval of h: here h is considered the 'lag', as in time series analysis. This gives the irregular line.

Then a smooth mathematical curve is fitted to this, as in regression, to give the Model Variogram (Figure 102) and this is used to calculate the weights in the weighted average procedure described previously. (The terms 'variogram' and 'semivariogram' are sometimes used interchangeably.)

For ordinary Kriging the assumption is that the mean μh is constant over the map ($\mu h = \mu$), but unknown, while the variogram γx, $x + h$ is known. Thus for each point p to be interpolated we need to find μ as well as each of the weights λi – where we know the distance h between the data point i and the interpolation point, and the sum of the weights equals 1. This gives the set of equations:

$$
\begin{vmatrix} \lambda 1 \\ \lambda 2 \\ \cdots \\ \lambda n \\ \mu \end{vmatrix}
\begin{vmatrix} \gamma 1,1 & \gamma 1,2 & \cdots & \gamma 1,n & 1 \\ \gamma 2,1 & \gamma 2,2 & \cdots & \gamma 2,n & 1 \\ \cdots & \cdots & \cdots & \cdots & 1 \\ \gamma n,1 & \gamma n,2 & \cdots & \gamma n,n & 1 \\ 1 & 1 & \cdots & 1 & 0 \end{vmatrix}
=
\begin{vmatrix} \gamma 1,p \\ \gamma 2,p \\ \cdots \\ \gamma n,p \\ 1 \end{vmatrix}
$$

which can be solved to find $\lambda 1$ to λn, as well as μ. The estimate of Zp, the elevation at the interpolation point p, is then the sum of $\lambda 1 Z 1 + \lambda 2 Z 2 + \cdots + \lambda n Z n$. Clearly the solution of the set of simultaneous equations for all interpolation points, with potentially quite a large number of neighbouring data points each, can be slow.

The ordinary Kriging error variance is calculated as $\lambda 1 \, \gamma 1,p + \lambda 2 \, \gamma 2,p + \cdots + \lambda n \, \gamma n,p + \mu$.

The great strength of Kriging is this estimate of the error, if all the conditions are true. However, according to Cressie (1993) Kriging might be bad if the assumptions are not true, or if the model variogram is not correct. If the nugget effect (the intersection of the variogram with the Y axis) is zero, then a map of the error variance will have zero values at each data point and increasing values in between.

In order to find the values $\gamma 1,1$, $\gamma 1,2$, etc. we must be able to have a model variogram to fit the experimental one. Common models are: circular; exponential; Gaussian; and linear – geostatistical packages have functions to fit these models to an experimental variogram. Several terms are worth remembering. The 'range' of the model is the distance h at which the variogram does not change significantly any more – points further away have no meaningful influence – a bit like a 'counting circle', while the 'sill' is the variance value at that point (see Figure 102). The 'nugget' is the variance at the origin: if this is not zero then in theory repeated sampling at the same location would give different results – this may be due to measurement errors or surface variation smaller than the sampling interval. For high-precision surveys, for example, a significant nugget effect would be worrying, but for samples of soil chemistry it would be expected.

As stated above, one of the assumptions of ordinary Kriging is that the mean μ for a selection of nearby points remains constant throughout the map. Where this is clearly not true then some regional 'trend' or 'drift' must be removed first, and the analysis done on the residuals – not forgetting to put the drift value back into the results afterwards. The drift is often estimated by simple bivariate regression. The variogram of the residuals should be examined closely: if the drift is not appropriate – for example a cubic polynomial was used to estimate an expected 'hump' in the terrain, but this is not really there – then the residuals have artificially-induced components, the mean differs significantly over the map, and the variogram shows no useful pattern.

The variogram may serve as an aid in the selection of the neighbouring data points used for the final interpolation. In Figure 102, for example, it is clear that beyond a certain distance – the range – the addition of further, more distant, points will not change the variation significantly. This helps in determining the radius of the 'counting circle' often used to select the set of neighbouring points used in the weighted average – excessively large circles will greatly increase the computation time. Care must be exercised, however: as in all weighted average methods, if the data distribution is very anisotropic then an insufficient number of data points will give a highly biased estimate.

Kriging can thus be seen as a rather complex method of interpolation: appropriate for highly variable data types such as soil chemistry but difficult to use, and it may not be worth the trouble for relatively precise survey data where the error is small – and perhaps the data distribution is a problem.

Summing up, the stages are:

a) Collect a subset of the data if the map is large;
b) Plot the Experimental Variogram;
c) Examine it to see if its behaviour is appropriate for the data type;
d) If necessary calculate a polynomial drift function, and subtract it from the data;
e) Examine the resulting Experimental Variogram to see if it is improved;
f) If so, select an appropriate Model Variogram from the available alternatives, and fit it to the experimental one;
g) For each interpolated point, solve the set of simultaneous equations to find the weights and mean;
h) Calculate the weighted average value of the interpolated point, and its error.

In conclusion, Kriging is appropriate for highly variable data, but may be unnecessary for more precise observations, where an error estimate is not required. Preliminary experiments suggest that in this case Sibson interpolation is faster, and just as good.

FURTHER READING

Kriging is of particular interest where the accuracy of the data being interpolated is known to be low. As described above, the technique departs significantly from the Voronoi approach used in this book – its variogram describing variability as a function of metric distance from the data point is opposed to the relative distances used in Voronoi methods. (Remember Newton vs. Leibniz described earlier: the argument has not disappeared!)

Matheron (1971) developed the original theory of regionalized variables, but his work is not easy reading. Olea (1984) provides a more reasonable summary, especially as applied to geological data, where it is still of significance. Oliver and Webster (1990) give an introduction for applications in the mapping of soil salinity – itself a highly variable property where simple interpolation procedures would not produce useful results. Gold and Condal (1995) found that Sibson interpolation produced more plausible surfaces for anisotropically distributed data such as the tracks of ship's soundings. Cressie (1993) is an appropriate reference.

3.7   MOVING POINTS: THE KINETIC ALGORITHM

The 'Dynamic' algorithm allows for point deletion as well as insertion, on a local basis, allowing the ongoing maintenance of the DT/VD graphs using only local updates. The same term is used for index graphs – trees – such as those used to store telephone numbers, etc. – the storage structure and algorithm are dynamic if individual telephone number records may be inserted or deleted without rebuilding everything. This is the basis for the insertion/deletion algorithm used for interpolation.

However, for spatial information – information for which location information is fundamental to its interpretation – we also have the need for point movement over time, and for this to be useful we need to maintain the connecting DT/VD structure as one or more points move. Here we not only need the graph structure at the point's initial and final positions – which we could achieve by deletion and insertion – but also all the possible intermediate structures as the point moves.

Initially, however, and for small perturbations, point movement may only make small changes to the geometry (coordinates) (Figure 103), and no changes to the graph structure itself.

However, when a point moves a greater distance the network must change to accommodate this: the 'empty circumcircle' criterion for a valid Delaunay triangulation will be broken and, as for the simple insertion algorithm described previously, pairs of triangles must have their diagonals switched to accommodate the change. This gives the basis of the moving-point algorithm, which is needed to keep track of the point movement.

In Figure 104a, point P is moving to the left. Initially the Voronoi cells P and Q are not neighbours – but the cells above and below are, meaning that a Delaunay edge connects them. As P moves to the

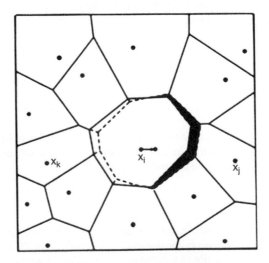

Figure 103.   Perturbing a data point.

78

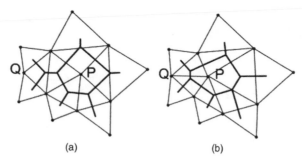

(a)   (b)

Figure 104.   A 'topological event'.

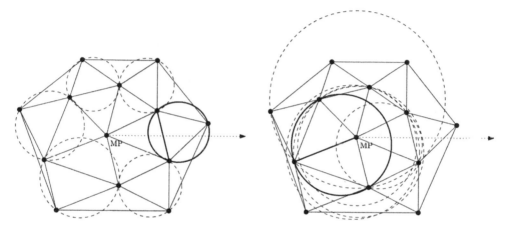

Figure 105.   Moving point entering and leaving a circumcircle.

left, as in Figure 104b, a moment is reached when these two cells are no longer neighbours, but P and Q are – which means that the Delaunay edge is switched to connect P and Q.

The key notion here is the idea of the 'topological event'. If one point in the DT is perturbed slightly, then its Voronoi boundaries change slightly also. However, if the displacement is large enough then one boundary will disappear, to be replaced by another. This is equivalent to the 'switch' of a triangle edge described earlier in the incremental construction of the VD. This takes place whenever the moving point moves into the previously-empty circumcircle of a 'real' Delaunay triangle – or moves out of the 'imaginary' circumcircle of a potential Delaunay triangle (formed by adjacent triples of points around P) which must then be re-formed as a triangle. If we can detect when the point moves into or out of these circumcircles, then we can maintain the VD/DT topological structure.

Figure 105 (left) shows the moving point MP in the process of moving to the right. It has six Delaunay neighbouring 'real' triangles and six empty circumcircles. As MP moves it could possibly enter any one of them first – making the circle and triangle invalid – and forcing a switch of the diagonal between the real triangle and the 'exterior' one. Thus the six circles need to be tested to see which of them MP would penetrate.

(A point needs to be made here: the intersection of a line with a circle unfortunately cannot be made with some variant of the CCW predicate – it is a quadratic with two solutions – and will have some small arithmetic error. The solution is always to use the earliest intersection, as this guarantees that all intersections will be used in the end. This then requires that 'previous' intersections, that are known to have already been passed, are flagged in advance and not tested – otherwise MP would move backwards!)

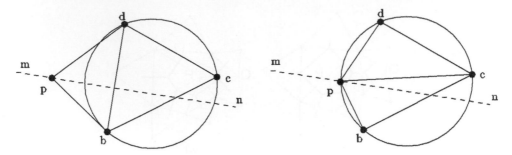

Figure 106.   Diagonal switching when a point enters a circumcircle.

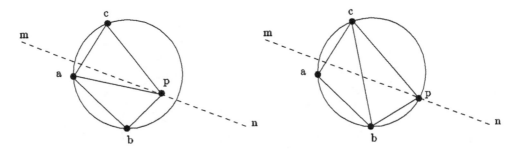

Figure 107.   Diagonal switching when a point leaves a circumcircle.

However, if MP moves into this circle it could quite possibly move out of it later; if it moves far enough the original triangle would have to be re-formed. Thus, in this case, the six 'ears' – triples of adjacent neighbouring points – need to be checked to see if they could possibly form a triangle as MP moves away. The 'earliest' of these intersections that lets MP exit a circle is chosen, and the diagonal switched – see Figure 105 (right).

Figure 106 shows point P entering a real circle: first the intersection is detected, then P is moved there, and then the diagonal is switched.

One extra test needs to be made. It is possible that point P, on the left, has a trajectory that enters the circle – but leaves it again between the same pair of points. In this case the circle is ignored. This is a simple line intersection test, as described previously, between P-n and, in this case, b-d. In this 'PAM' test (Point-Arc Matching, but named after my daughter who devised it – Gold, (1992)), a positive intersection means that the diagonal must be switched: a negative result means no switch is needed.

Figure 107 shows point P leaving an imaginary circle: first the exit intersection of the trajectory with the circle is detected, then P is moved there, and then the diagonal is switched.

If we can detect when the point moves into or out of these circumcircles, then we can maintain the VD/DT topological structure, as illustrated in Figure 108 to Figure 114.

As shown above, by keeping track of the moving point's entry into the next 'real' circumcircle, and exit from the next 'imaginary' one, the Delaunay property of the triangulation may be maintained – and thus the Voronoi diagram is maintained also.

A remaining issue is collision, as multiple Voronoi generators may not occupy the same location. With simple insertion, if a point is inserted or moves too close to another, as specified by a tolerance, then it is ignored. With a moving point in a VD, if it moves too close to another point it is merged with it. Depending on the application, it may be recreated (split off) again to continue its trajectory.

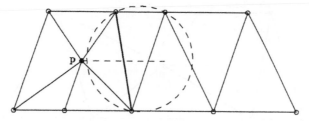

Figure 108. Maintaining the DT: stage 1.

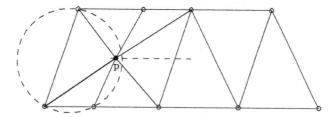

Figure 109. Maintaining the DT: stage 2.

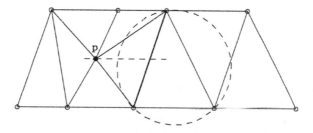

Figure 110. Maintaining the DT: stage 3.

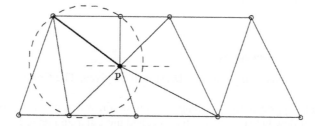

Figure 111. Maintaining the DT: stage 4.

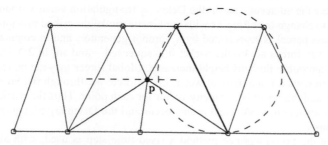

Figure 112. Maintaining the DT: stage 5.

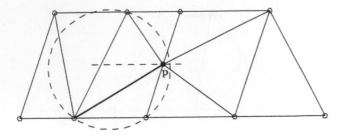

Figure 113.   Maintaining the DT: stage 6.

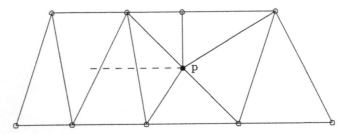

Figure 114.   Maintaining the DT: stage 7.

FURTHER READING

Within computational geometry Guibas et al. (1991), Fu and Lee (1991) and Roos (1993) were concerned with maintaining the mesh for many simultaneously-moving points: predicting the next 'topological event', based on solving quartic equations. Within the GIS community, Gold (1990b) looked at triangle edge flipping to manage moving points, and Mostafavi and Gold (2004), mentioned previously, examined moving points on the sphere moving one point at a time, the earliest topological event first.

3.7.1   Free-Lagrange flow modelling

Two important applications of the kinetic VD/DT will be given. The first is Free-Lagrange flow modelling.

The most common approach to modelling the flow of air, water, etc. is to fix the buckets and model the amount of flow between them – this is called the Eulerian method. The alternative is to fix the amount of fluid in each parcel and monitor the flow-line of the parcel over time – this can be useful, for example, where the paths of tracers need to be modelled. The term 'Free-Lagrange' is usually used where the adjacency mesh (the Delaunay triangulation when Voronoi cells are being used) is allowed to change over time, as with our kinetic model described previously.

Each fixed-mass parcel is represented by a Voronoi generator, and is contained in a Voronoi cell 'bag', so that as forces act on the packet it is squeezed – and so, in 2D, its area is reduced and, for an incompressible fluid, its height increases. Global forces may act on each particle (data point), which represents a fixed mass of water. Local forces act through the Voronoi neighbours, jostling together, and the result is a velocity vector for each point. Thus the particle moves and, if the movement is sufficient, topological events occur and the Delaunay triangulation changes, as described previously.

Normally with the Free-Lagrange method a fixed time step is used, with all the forces and movements recalculated for each interval, as with finite-difference or finite-element methods. This

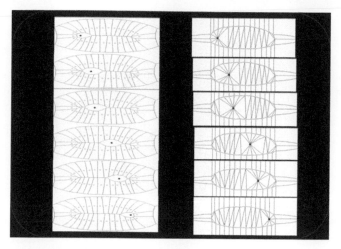

Figure 115.   A moving point in a VD, and a DT.

Voronoi diagram

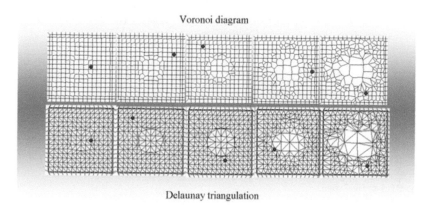

Delaunay triangulation

Figure 116.   A 'dam breaking' simulation.

method can cause problems, as one fast-moving particle may collide with another in a single time step, or else overshoot it, destroying the mesh structure.

This problem may be eliminating by removing the fixed time step, and replacing it with a priority-queue of the topological events. Each moving point has its next topological event, when a triangle switch must take place, and the time of this depends on the particle's velocity. Thus at any stage a queue is preserved of the next topological event for each particle, and the earliest of these events is then processed: the particle is moved, the edge switched, and the effect of the new neighbour on the moving particle (and vice versa) is calculated and added to the total force, and motion, of each particle. (The same process is performed to remove the effect of old neighbours just removed.) The timing of the next topological event is then recalculated for these particles, and the priority queue updated. This process is then repeated, with whatever time interval is necessary to arrive at the next event. Figure 115 shows the VD and DT for a single particle travelling through a mesh.

Figure 116 shows a traditional 'dam breaking' exercise: imagine that in the central region the water is at a higher level in a box. This box is then suddenly removed, and the waves splash outwards.

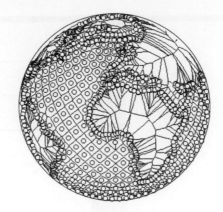

Figure 117.   Tides: initial state.

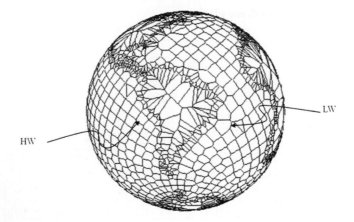

Figure 118.   Tides: final state.

Note that on the far right the Voronoi cells are considerably larger, indicating that the water level is lower than when the process started – as would be the experimental situation.

In one study, global tides were modelled: first the Voronoi cells were calculated on the sphere, as described previously. For the tidal modelling, fixed points were added along the coastlines, and regular Voronoi generators representing fixed-mass packets of water were added within the oceans (Figure 117).

Then the Free-Lagrange process was started, with local neighbouring forces plus lunar gravity impinging on each packet. As can be seen in Figure 118, the smaller cells (high water) are here at the west coast of South America, and larger cells (low water) on the east coast. (This model did not include the earth's rotation.)

Almost the whole globe is shown in Figure 119, in the Mercator projection. The flow vectors of each packet are also indicated.

It should be noted that the coastlines here are really double-point boundaries, as described later. "By definition, each Voronoi cell represents the spatial extension of the particle that generates it and the Voronoi edge between two adjacent particles may change if one of the particles moves. Therefore it is not appropriate to represent the interface between two different materials with Voronoi edges. Using two parallel sets of stationary particles ensures the precise representations of the coasts," Mostafavi and Gold (2004). This means that the inner coastal set will not change, but the outer ocean set may have their seaward boundaries modified as the water level changes.

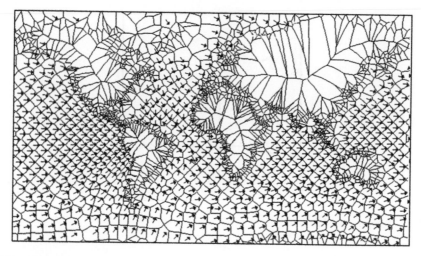

Figure 119.   Mercator projection.

FURTHER READING

The alternative to an Eulerian (fixed-mesh) model for fluid flow simulation is a Lagrangian (fixed cell volume) approach. Early approaches to this encountered 'mesh tangling', when one cell ended up moving past its neighbours. The free-Lagrange approach takes advantage of kinetic meshes – usually Delaunay – to manage this. The first widely known work on this was the book by Fritts et al. (1985), with entries by Augenbaum on a Voronoi mesh on the sphere, and Eltgroth on using independent, instead of fixed, time steps. Later work was by Crowley (1988), Trease et al. (1990), Rees and Morton (1991) and Ball (1996). All of these, and others, were concerned with preserving the fluid-flow equations.

3.7.2   Kinetic GIS – The marine GIS

A second important application of the kinetic VD/DT is the development of a Marine GIS.

The kinetic VD allows one of the basics of navigation – collision detection – as the examination of a boat's immediate neighbours at any point in time will give those most likely to be hit if the current course is continued. These neighbouring points could also be the outlines of islands, wrecks or other obstacles, as in Figure 120, an early sketch from Gold and Condal (1995). 'A' is a surface vessel navigating between the shore and the island.

We can also perform channel navigation with our moving boat, considering the neighbouring points to be sea bottom observations, by using Sibson interpolation – 'B' in Figure 120 is a vessel estimating the depth from nearby bathymetry. Thus at any point in time our 'ship' will have a knowledge of the nearest surface objects – other ships and coastlines – as well as being able to estimate its depth by interpolating from its neighbouring bathymetric soundings. This led to the development of a ship navigation system based on the kinetic VD, along with a 3D cartographic view of the area – viewed from any overhead position, from the ship's bridge, or even from below the surface. Figure 121 shows the Voronoi diagram.

Figure 122 is a view from the ship's bridge – note the sea-lane chart markings superimposed on the water surface. Ship positions are taken in real time from on-board AIS beacons.

This application was considered worth developing commercially (Figure 123, Figure 124, Figure 125): for further details see **http://www.srt-marinesystems.com/**

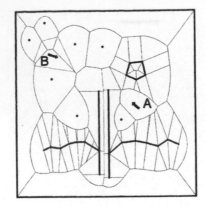

Figure 120.   Early marine simulation.

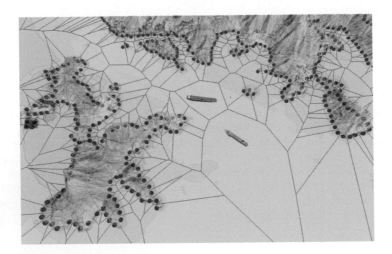

Figure 121.   VD for marine simulation.

Figure 122.   View from the ship's bridge.

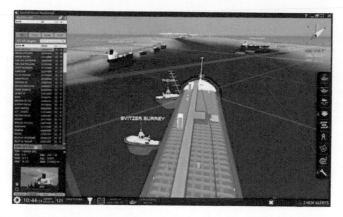

Figure 123.   Bridge view.

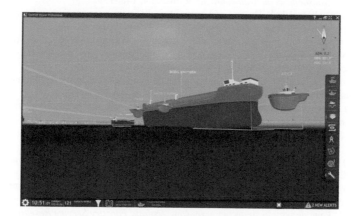

Figure 124.   Subsea view.

Figure 125.   Harbour 3D view.

FURTHER READING

Another major application was the development of the 'Marine GIS'. Early work discussing the subject in general included Davis and Davis (1998), Li and Saxena (1993), Lockwood and Li (1995), Goulielmos and Tzannatos (1997) and Wright and Goodchild (1997) – as well as various papers in Wright and Bartlett (2000). The use of kinetic data structures was first discussed in Gold and Condal (1995), mentioned previously, then Gold (2000). Full 3D graphics for marine navigation was outlined in Gold et al. (2004), Gold and Goralski (2007), Goralski and Gold (2008), Goralski et al. (2011), based on the International Hydrographic Organization (1996) chart specification. This has resulted in the current commercial GeoVS maritime awareness system marketed by SRT plc. Collision avoidance is managed by the kinetic VD/DT, and bathymetric prediction by Sibson interpolation.

REFERENCES

Akima, H. (1978) A method of bivariate interpolation and smooth surface fitting for irregularly distributed data points. *ACM Transactions on Mathematical Software*, 4, 148–159.

Arnold, D.B. & Milne, W.J. (1984) The use of Voronoi tessellations in processing soil survey results. *IEEE Computer Graphics and Applications*, 4, 22–28.

Augenbaum, J.M. (1985) A Lagrangian method for the shallow water equations based on Voronoi mesh-flows on a rotating sphere. In: Fritts, M.J., Crowley, W.P. & Trease, H. (eds.) *The Free-Lagrange Method, Lecture Notes in Physics*. Vol. 238. Berlin, Springer-Verlag. pp. 54–86.

Bajaj, C.L. & Bouma, W.J. (1990) Dynamic Voronoi diagrams and Delaunay triangulations. In: *Abstracts of the Second Canadian Conference on Computational Geometry*. pp. 273–277.

Ball, G.J. (1996) A Free-Lagrangian method for unsteady compressible flow: Simulation of a confined cylindrical blast wave. *Shock Waves*, 5, 311–325.

Bengtsson, B.-E. & Nordbeck, S. (1964) Construction of isarithm and isarithmic maps by computer. *BIT*, 4, 87–105.

Boas, S.W. (1951) Delimitation of seaward areas under national jurisdiction. *American Journal of International Law*, 45, 240–266.

Boissonnat, J.D. & Cazals, F. (2002) Smooth surface reconstruction via natural neighbour interpolation of distance functions. *Computational Geometry – Theory and Applications*, 22, 185–203.

Brown, G.S. (1965) Point density in stems per acre. *New Zealand Forestry Service Research Notes*, 38, 1–11.

Caulfield, D.D., Kenway, D.J. & Gold, C.M. (1984) Stereo side-scan as a complement to echo-sounding for high resolution bathymetric studies. In: *Report, USN/SEG Three Dimensional Marine Data Collection, Processing, Interpretation and Presentation: Fourth Biennial Society of Exploration Geophysics/U.S. Navy Joint Technical Symposium, Diamond Head, MS, USA*.

Cressie, N. (1993) *Statistics for Spatial Data*. New York, Wiley Interscience.

Crowley, W.P. (1988) A Free-Lagrange method for 2-D compressible flows. *Computer Physics Communications*, 48, 51–60.

Davis, B.E. & Davis, P.E. (1988) Marine GIS: Concepts and considerations. In: *Proceedings, GIS/LIS '88*.

Devillers, O. (1999) On deletion in Delaunay triangulations. In: *Proceedings, 15th Annual ACM Symposium on Computational Geometry*. pp. 181–188.

Devillers, O. (2002) On deletion in Delaunay triangulations. *International Journal of Computational Geometry and Applications*, 12, 193–205.

Eltgroth, P.G. (1985) Free Lagrange-Methods, independent time steps, and parallel processing. In: Fritts, M.J., Crowley, W.P. & Trease, H. (eds.) *The Free-Lagrange Method, Lecture Notes in Physics*. Vol. 238. Berlin, Springer-Verlag. pp. 114–122.

Farin, G. (1990) Surfaces over Dirichlet tessellations. *Computer Aided Geometric Design*, 7, 281–292.

Fritts, M.J., Crowley, W.P. & Trease, H. (1985) *The Free-Lagrange Method. Lecture Notes in Physics*. Vol. 238. Berlin, Springer-Verlag.

Fu, J.-J. & Lee, R.C.T. (1991) Voronoi diagrams of moving points in the plane. *International Journal of Computational Geometry and Applications*, 1, 23–32.

Gold, C.M. (1979) Triangulation based terrain modelling – Where are we now? In: Aangeenburg, R.T. (ed.) *Proceedings, Auto-Carto 4, International Symposium on Cartography and Computing*. Vol. 2. pp. 104–111.

Gold, C.M. (1987) Spatial ordering of Voronoi networks and their use in terrain data base management. In: *Proceedings, Auto-Carto*. Vol. 8. pp. 185–194.

Gold, C.M. (1988) Point and area interpolation and the digital terrain model (with discussion). In: Lee, Y.C. (ed.) *Proceedings, Trends and Concerns of Spatial Sciences, Second Annual International Symposium of the Atlantic Institute*. New Brunswick, University of New Brunswick. pp. 133–147.

Gold, C.M. (1989) Chapter 3 – Surface interpolation, spatial adjacency and G.I.S. In: Raper, J. (ed.) *Three Dimensional Applications in Geographic Information Systems*. London, Taylor & Francis. pp. 21–35.

Gold, C.M. (1990a) Neighbours, adjacency and theft – The Voronoi process for spatial analysis. In: Ottens, H.F.L., Harts, J. & Scholten, H.J. (eds.) *Proceedings First European Conference on Geographic Information Systems*. Amsterdam, The Netherlands. pp. 382–391.

Gold, C.M. (1990b) Spatial data structures – The extension from one to two dimensions. In: Pau, L.F. (ed.) *Mapping and Spatial Modelling for Navigation*. NATO ASI Series F No. 65. Berlin, Springer-Verlag. pp. 11–39.

Gold, C.M. (1994) An object-based method for modelling geological surfaces containing linear data. In: *Proceedings, Annual Meeting of the International Association for Mathematical Geology*. Mont Tremblant, QC, Canada. pp. 141–146.

Gold, C.M. (2000) An algorithmic approach to marine GIS. In: Wright, D.J. & Bartlett, D. (eds.) *Marine and Coastal Geographical Information Systems*. London, Taylor & Francis. pp. 37–52.

Gold, C.M. & Condal, A.R. (1995) A spatial data structure integrating GIS and simulation in a marine environment. *Marine Geodesy*, 18, 213–228.

Gold, C.M. & Goralski, R. (2007) 3D graphics applied to maritime safety. In: Popovich, V.V., Schrenk, M. & Korolenko, K.V. (eds.) *Information Fusion and Geographic Information Systems*. Berlin, Springer. pp. 286–300.

Gold, C.M., Charters, T.D. & Ramsden, J. (1977) Automated contour mapping using triangular element data structures and an interpolant over each triangular domain. In: George, J. (ed.) *Proceedings Siggraph '77. Computer Graphics*. Vol. 11. pp. 170–175.

Gold, C.M., Chau, M., Dzieszko, M. & Goralski, R. (2004) 3D geographic visualization: The marine GIS. In: Fisher, P.F. (ed.) *Developments in Spatial Data Handling – 11th International Symposium on Spatial Data Handling*. Berlin, Springer. pp. 17–28.

Gold, P.D. (1992) Personal communication.

Goralski, R. & Gold, C.M. (2008) Marine GIS: Progress in 3D visualization for dynamic GIS. In: Ruas, A. & Gold, C.M. (eds.) *Headway in Spatial Data Handling*. Berlin, Springer. pp. 401–416.

Goralski, R., Ray, C. & Gold, C.M. (2011) Applications and benefits for the development of cartographic 3D visualization systems in support of maritime safety. In: Weintrit, A. (ed.) *International Recent Issues About ECDIS, e-Navigation and Safety at Sea*. Boca Raton, FL, CRC Press. pp. 77–86.

Goulielmos, A. & Tzannatos, E. (1997) Management information system for the promotion of safety in shipping. *Journal of Disaster Prevention and Management*, 6, 252–262.

Green, P.J. & Sibson, R. (1978) Computing Dirichlet tessellations in the plane. *The Computer Journal*, 21, 168–173.

Guibas, L.J., Mitchell, J.S.B. & Roos, T. (1991) Voronoi diagrams of moving points in the plane. In: *Proc. 17th Internat. Workshop on Graph-Theoretical. Concepts in Computer Science, Lecture Notes in Computer Science*. Vol. 570. Berlin, Springer-Verlag. pp. 113–125.

Hatcher, G.A.J. & Maher, N. (2000) Real-time GIS for marine applications. In: Wright, D.J. & Bartlett, D. (eds.) *Marine and Coastal Geographical Information Systems*. London, Taylor & Francis. pp. 137–147.

Heller, M. (1990) Triangulation algorithms for adaptive terrain modeling. In: *Proceedings 4th International Symposium on Spatial Data Handling*. pp. 163–174.

International Hydrographic Organization (1996) *Specifications for Chart Content and Display Aspects of ECDIS*. (5th edition, December 1996, amended March 1999 and December 2001) – Special Publication No. 52. Monaco, International Hydrographic Bureau.

Johnson, W.A. & Mehl, R.F. (1939) Reaction kinetics in processes of nucleation and growth. *Transactions of the American Institute of Mining, Metallurgical and Petroleum Engineers*, 135, 410–458.

Lam, N.S.-N. (1983) Spatial interpolation methods: A review. *The American Cartographer*, 10, 129–149.

Lawson, C.L. (1977) Software for C^1 surface interpolation. In: Rice, J. (ed.) *Mathematical Software III*. New York, NY, Academic Press. pp. 161–194.

Lee, J. (1991) Comparison of existing methods for building triangular irregular network models. *International Journal of Geographical Information Systems*, 5, 267–285.

Li, R. & Saxena, N.K. (1993) Development of an integrated marine geographic information system. *Journal of Marine Geodesy*, 16, 293–307.

Li, Z., Zhu, Q. & Gold, C.M. (2005) *Digital Terrain Modelling – Principles and Methodology*. London, CRC Press.

Lockwood, M. & Li, R. (1995) Marine geographic information systems – What sets them apart? *Marine Geodesy*, 18, 157–159.

Lowell, K. (1994) A fuzzy surface cartographic representation for forestry based on Voronoi diagram area stealing. *Canadian Journal of Forest Research*, 24, 1970–1980.

Matheron, G. (1971) The theory of regionalised variables and its applications. *Les Cahiers du Centre de Morphologie Mathématique de Fontainebleu* 5.

Mead, R. (1967) A mathematical model for the estimation of inter-plant competition. *Biometrics*, 23, 189–205.

Mostafavi, M.A., Gold, C.M. & Dakowicz, M. (2003) Delete and insert operations in Voronoi/Delaunay methods and applications. *Computers & Geosciences*, 29, 523–530.

Olea, R.A. (1984) Systematic sampling of spatial functions. *Kansas Geological Survey Series on Spatial Analysis* No. 7.

Oliver, M.A. & Webster, R. (1990) Kriging: A method of interpolation for geographical information systems. *International Journal of Geographical Information Systems*, 4, 313–332.

Pelz, D.R. (1978) Estimating individual tree growth with tree polygons. In: Freis, J., Burkhart, H.E. & Max, T.A. (eds.) *Growth Models for Long-Term Forecasting of Timber Yields*. Virginia Polytechnical Institute and State University, School of Forestry and Wildlife Resources, Publications FWS-1-78. pp. 172–178.

Peucker, T.K. & Chrisman, N. (1975) Cartographic data structures. *The American Cartographer*, 2, 55–69.

Peucker, T.K., Fowler, R.J., Little, J.J. & Mark, D.M. (1978) The triangulated irregular network. In: *Proceedings of the Digital Terrain Models Symposium*. St. Louis, American Society of Photogrammetry. pp. 16–54.

Philip, G.M. & Watson, D.F. (1982) A precise method for determining contoured surfaces. *Journal of the Australian Petroleum and Exploration Association*, 22, 205–212.

Rees, M.D. & Morton, K.W. (1991) Moving point, particle, and Free-Lagrange methods for convection-diffusion equations. *SIAM Journal on Scientific and Statistical Computing*, 12, 547–572.

Rhind, D. (1975) A skeletal overview of spatial interpolation techniques. *Computer Applications*, 2, 293–309.

Ricketts, P. (1986) Geography and international law: The case of the 1984 Gulf of Maine boundary dispute. *The Canadian Geographer*, 30, 194–205.

Roos, T. (1993) Voronoi diagrams over dynamic scenes. *Discrete Applied Mathematics*, 43, 243–259.

Saito, N. (1982) Asymptotic regular pattern of epidermal cells in mammalian skin. *Journal of Theoretical Biology*, 95, 559–599.

Sambridge, M., Braun, J. & McQueen, H. (1995) Geophysical parameterization and interpolation of irregular data using natural neighbours. *Geophysical Journal International*, 122, 837–857.

Shamos, M.I. (1975) Geometric complexity. In: *Proc. 7th. Ann. Symp. Theory of Computing*. pp. 224–233.

Sibson, R. (1978) Locally equiangular triangulations. *The Computer Journal*, 21, 243–245.

Sibson, R. (1980) A vector identity for the Dirichlet tessellation. *Mathematical Proceedings of the Cambridge Philosophical Society*, 87, 151–155.

Sibson, R. (1981) A brief description of natural neighbour interpolation. In: Barnett, V. (ed.) *Interpreting Multivariate Data*. New York, NY, John Wiley. pp. 21–36.

Sibson, R. (1982) A brief description of natural neighbour interpolation. In: Barnett, V. (ed.) *Interpreting Multivariate Data*. London, John Wiley and Sons. pp. 21–36.

Smalley, I.J. (1966) Contraction crack networks in basalt flows. *Geological Magazine*, 103, 110–114.

Su, Y. (2000) A user-friendly marine GIS for multi-dimensional visualisation. In: Wright, D.J. & Bartlett, D. (eds.) *Marine and Coastal Geographical Information Systems*. London, Taylor & Francis. pp. 227–236.

Thiessen, A.H. (1911) Precipitation averages for large areas. *Monthly Weather Review*, 39, 1082–1084.

Trease, H.E., Fritts, M.J. & Crowley, W.P. (1990) *Advances in the Free-Lagrange Method, Lecture Notes in Physics*. Vol. 395. Berlin, Springer-Verlag.

Watson, D.F. & Philip, G.M. (1987) Neighbourhood-based interpolation. *Geobyte*, 2, 12–16.

Wright, D.J. & Goodchild, M.F. (1997) Data from the Deep: Implications for the GIS Community. *International Journal of Geographical Information Science*, 11, 523–528.

Chapter 4

Boundaries

4.1 CLASS I: LINES OF POINTS

4.1.1 *Crust and skeleton test*

Labelled skeletons may be used to separate points that have additional information available, but more often this is not the case. In particular, for strings of points, where the distance between consecutive points is clearly less than the distance separating strings, it is often desirable to 'connect the dots' (to form the 'crust') – and also to define skeletons, as before, composed of portions of Voronoi cell boundaries, that separate the strings. To achieve this we need to go back to before the availability of digital computers, when (Blum 1967) was attempting to describe 'biological shapes' – the following five sketches were derived from his work.

These were based on the idea of 'wave fronts' spreading out from the generators – which could be points or continuous lines (Figure 126, Figure 127). Where the wave fronts meet we have Voronoi boundaries.

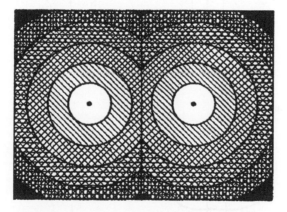

Figure 126.   Wave fronts meeting.

Figure 127.   Wave fronts for various generators.

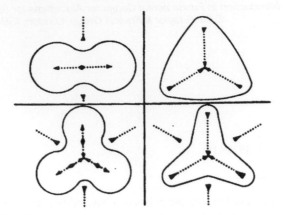

Figure 128.   Biological shapes.

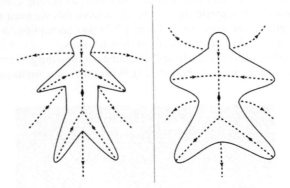

Figure 129.   Anthropomorphs.

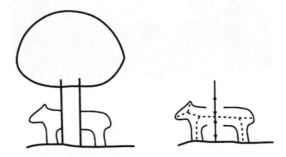

Figure 130.   Perceptual model.

The generators could be the boundaries of his biological shapes, which cannot be defined using simple line segments and arcs. These are usually called the 'medial axis', but here we call them the 'geometric skeleton', to distinguish it from the labelled skeleton discussed earlier (Figure 128).

The skeletons may have a form that suggests its 'shape' – here the possibility that the outline represents a person (Figure 129).

These forms may be similar to those generated by human visual processes, for example edge detection. They may have been useful in detecting animals – even predators. In Figure 130 two striped portions of the image are obscured by a tree – but the skeleton suggests the complete shape – a tiger!

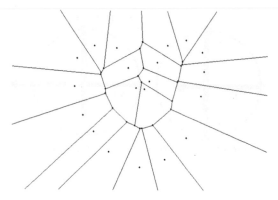

Figure 131.   VD of a closed curve.

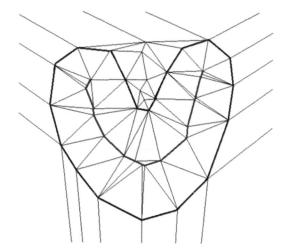

Figure 132.   Amenta's crust detection.

Crust and skeleton

One of the features of the Voronoi diagram is that it can be used to connect a set of points on an open or closed curve (the 'crust'), or to construct the medial axis of a polygon or open curve (the 'skeleton'). Figure 131 shows the VD of points along a closed curve: the Voronoi centres clearly mark the skeleton.

The work of Amenta et al. (1998) showed one way to determine the boundary of the curve: insert these Voronoi centres into the original diagram, and re-triangulate. All edges not connected to the Voronoi centres are part of the boundary (Figure 132). They also described a limit on how close together the boundary points must be in order to preserve the crust continuity: r, the 'Local Feature Size', must be less than 0.4, where r is the distance between some point p on the crust and the closest point on the medial axis. If this figure is exceeded then the crust connectivity may be lost. In practice this means that sharp corners, where r is very small, will probably lose crust continuity.

Later work by Gold (1999) simplified the process. In the original VD/DT, every Delaunay edge has a matching Voronoi edge. Using the previous defined circumcircle test 'H', construct the circle with the Delaunay vertices and one Voronoi vertex: if the other Voronoi vertex falls within the circle then the Voronoi edge is part of the skeleton and the Delaunay edge is discarded (left side of Figure 133); if it falls outside the circle then the Delaunay edge is part of the crust and the Voronoi

93

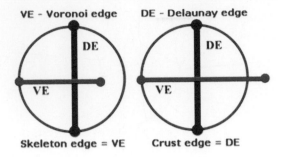

Figure 133. Crust/skeleton test.

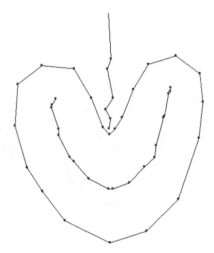

Figure 134. Crust and skeleton.

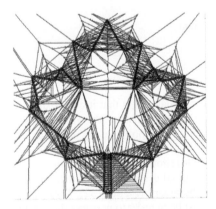

Figure 135. Crust and skeleton of a maple leaf.

edge is discarded (right side of Figure 133). Thus of every VD/DT edge pair, one is preserved and the other discarded.

The resulting heart-shaped figure is shown in Figure 134.

In a more interesting example the outline of a maple leaf was scanned and points generated along the boundary. The resulting VD/DT is shown in Figure 135. Clearly some Voronoi edges form the

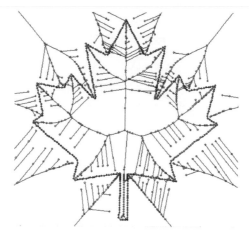

Figure 136. Maple leaf crust and skeleton.

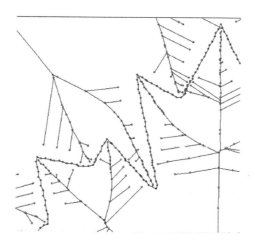

Figure 137. Skeleton 'hairs'.

skeleton and some (unwanted) ones pass between the boundary points. Similarly, some Delaunay edges form part of the crust, and some unwanted ones cross the skeleton.

After performing the above selection test, the crust and skeleton are clearly displayed (Figure 136). However, there are many 'hairs' on the skeleton, due to the irregularity of points along the crust. These can be removed by perturbing the crust boundary points appropriately.

Figure 137 shows some details of the original figure. Hairs are generated when triples of boundary points form nearly flat triangles, with Voronoi centres far away from the main skeleton.

The ends of these skeleton hairs may be 'retracted' to their parent skeleton points, as shown in Figure 138, by pulling the exterior point in until it lies on the parent triangle's circumcircle – this superimposes the two skeleton points and removes most hairs.

The test must be repeated on the other side of the boundary as well, which means the original circles are perturbed – meaning the process must be repeated, However, it converges very rapidly to a stable smoothed boundary.

The resulting maple leaf is shown in Figure 139. Note that there are interior (endo-) and exterior (exo-) skeletons produced. No tolerances are needed for this smoothing, removing a frequent source of difficulties.

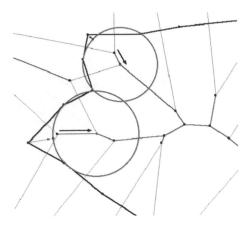

Figure 138.   Skeleton retraction.

Figure 139.   Final smoothed maple leaf.

FURTHER READING

Labelled skeletons may delimit proximal regions if points have distinctive labels. When this is not the case – and especially when we are looking at closely spaced lines of points – another approach must be used.

Blum (1967, 1973) and Blum and Nagel (1978) examined the intersections of wave-fronts generated by collections of points and defined the 'medial axis' or 'skeleton' at the mid points, and Christensen (1987) described the medial axis as the wave-front intersections from contour lines. This was developed prior to digital computer algorithms for the VD/DT. Gold (1992) described this approach, and Ogniewicz (1992) and Ogniewicz and Ilg (1992) described methods for pruning the Voronoi diagram to obtain the skeleton. Amenta et al. (1998) showed a method for extracting the 'crust' – the connected set of boundary points – of the polygon, and gave constraints on its validity. Gold (1999) and Gold and Snoeyink (2001) described the simple extraction of both the crust and the skeleton from the boundary points, as mentioned above.

Gold and Thibault (2001) showed a simple method for smoothing the often 'hairy' results of skeleton extraction, and Gold and Dakowicz (2005) described a variety of applications in GIS. Kimia (2004) returned to Blum's original interest in human visual perception.

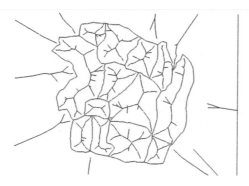

Figure 140.   Crusts and skeletons of a polygon set.

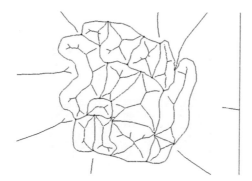

Figure 141.   Smoothed crusts and skeletons.

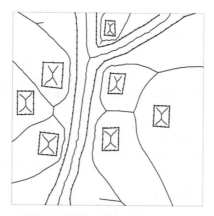

Figure 142.   Crusts and skeletons: a simple cartographic dataset.

4.1.2   *Polygon 'shapes'*

Crusts and skeletons may be formed for any polygon sets, as shown below – first unsmoothed (Figure 140), and then smoothed (Figure 141).

Crusts and skeletons may be generated from simple cartographic input, as shown in Figure 142. The crust outlines the digitized buildings and roads, and the skeleton gives the road centreline and the building form and 'territory' – the land closest to it.

Many other applications exist. In Figure 143, after Ogniewicz (1992), images of scissors were scanned on a production conveyor belt, and edge-detection filters used to identify the boundaries.

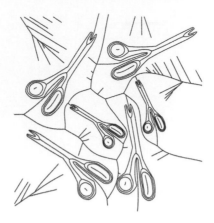

Figure 143.   Scanned images of scissors.

Figure 144.   VD/DT for three characters.

Crusts and skeletons were then generated. The endo-skeletons of the scissors give the form of the object, which can be interpreted, allowing a robot to pick up the particular item. The exoskeletons show the proximity of each item, which may be used to identify problems such as objects on top of each other.

Skeletons are also useful in character recognition. As shown in Figure 144, Figure 145 and Figure 146, the VD/DT of the boundaries are generated, and the crust and skeleton determined, and then smoothed. The form of the endoskeleton is of major help in the determining the individual letters – and the exoskeletons may be used to identify the order of characters to form words.

4.1.3   Contours – skeleton enrichment

One particularly interesting type of boundary 'shape' is that of contour lines representing a topographic surface. While direct LiDAR and similar data acquisition methods are increasingly used, there are many cases where it is necessary to use previously-drawn contour maps as data sources. As shown previously, traditional 'counting circle' interpolation has many problems with strings of data points, but the Voronoi-based Sibson approach can produce reasonable elevation estimates even close to the contour lines themselves.

One particular problem remains, as shown with the sample contour data set of Figure 86. While intermediate interpolated values are reasonable between contours of differing elevations, whenever we have ridges, valleys, summits or pits all nearby contour values are identical – producing a 'flat' region even when we know there will be a ridge or a valley (Figure 87). However, looking at the Voronoi cells and the skeleton, superimposed on the terrain, indicates a possible approach.

Figure 145.   Crusts and skeletons for the characters.

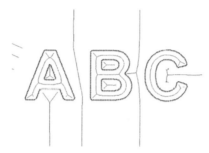

Figure 146.   Smoothed crusts and skeletons.

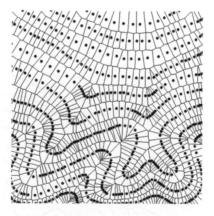

Figure 147.   Voronoi cells of the contour dataset.

In Figure 87 we can see that within ridges and valleys many Delaunay triangles connect to the same contour line, and thus all points within them have the same elevation. This is clearly incorrect, and visually distracting. In these cases the skeleton points – the circumcentres of these triangles – may be added to the TIN: this action breaks up the flat triangles, but they must still be assigned elevation estimates.

When we examine the Voronoi cells – the dual of the Delaunay triangulation – we can see ridge and valley lines (Figure 147) – and when we extract the skeleton (Figure 148) these are clearly shown, along with the more obvious intermediate skeletons between differing contour levels (Figure 149).

Figure 148.   Skeletons of the contour dataset.

Figure 149.   Skeletons superimposed on the terrain.

Figure 150.   Residual skeletons.

Removing the intermediate skeleton portions leaves us with the remainder (Figure 150): new simulated data points to enrich the form of the ridges and valleys. These break up the original 'flat' triangles. Nevertheless, reasonable elevation estimates need to be made at these locations in order to provide meaningful ridge and valley lines. This involves examining the medial axis properties: points on the skeleton are formed by Voronoi cell vertices, and each vertex is the centre of the circumcircle of the triangle containing the vertex.

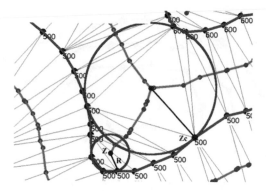

Figure 151.   Estimating ridge and valley elevations.

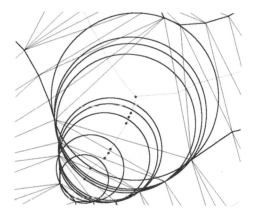

Figure 152.   Skeleton point circumcircles.

4.1.4   *Contours – skeleton point elevation*

The circumcircles may be used to estimate the elevations, following the assumption of equal slope for all points within the ridge or valley being evaluated: lacking any other information it can be postulated that valley walls are of constant slope, being a function of rock type, erosion forces, etc. So, if we know the expected slope, then the elevation of the skeleton point can be calculated.

The expected slope may be estimated from the reference circumcircle at the junction of the minor ridge or valley with its parent: the junction is known to be half way between two contours of different elevation: its own elevation is thus available, the circle radius to the adjacent contours is known, and thus the ridge or valley slope may be modelled. At any intermediate point on the skeleton branch the radius is also known, as is the slope, and so the elevation may be estimated – see Figure 151.

Here the large circle (with its slope estimate) and the smallest circle, at the far end of the valley, are known. The ratio of the two radii gives the ratio of the height difference between the skeleton point and the circumscribing contour.

Figure 152 shows the circumcircles for all the skeleton points of the small valley.

The same approach may be used for summits or pits (Figure 153) – but here the initial slope estimate must be calculated a little differently. First select a terminal vertex of the skeleton, obtain its radius, and look at its 'unused' Voronoi edges that cross the crust. Then examine the far Voronoi vertex – it will be between two differing levels, and so it will have both a radius and an elevation. This is the reference circle, which is then used exactly as for valleys. (The selection of different

101

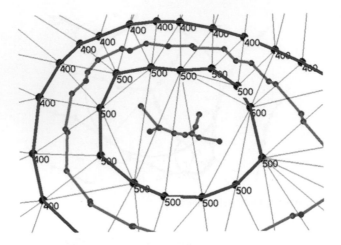

Figure 153.   Skeleton at a summit or pit.

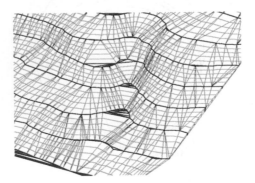

Figure 154.   Down-slope stream: flat triangles.

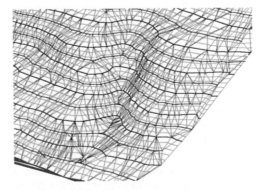

Figure 155.   Down-slope stream: added skeleton points.

reference circles may change the calculated summit values slightly, but in practice the differences are small.)

A real-world example is shown, first with the flat triangles visible (Figure 154) and then with the skeleton vertices and elevations added (Figure 155).

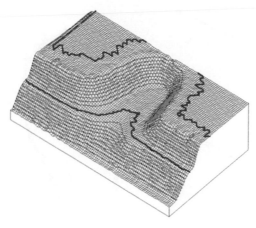

Figure 156.   Hanging valley: flat triangles.

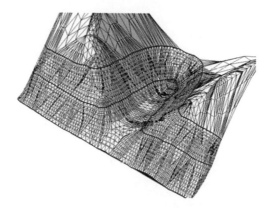

Figure 157.   Hanging valley: skeleton enrichment.

A synthetic example shows the limitations of the method: a simple hanging valley sits above a steeper main valley. A traditional interpolation gives a flat valley floor (Figure 156), while the skeleton enrichment produces a plausible valley (Figure 157): whether it should drop off so steeply to the main valley is unclear.

As a side note, it is possible to estimate skeleton point heights from the ratios of the valley lengths, as shown in Figure 158, but the results were not as good as the constant-slope method.

Skeleton retraction – with the concomitant crust modification – is a fairly general procedure, and not restricted exclusively to the first order hair retraction, where the ends of the hairs are pulled in to their immediate parent point on the skeleton. This, by its nature, limits the extent by which the crust may be simplified. However, other retraction parameters may be used, perhaps simplifying the crust to a greater extent than this. One example is the work of Ogniewicz and Ilg (1992), mentioned previously, who examined the cases of retraction based on the boundary length between two points on the crust (perpendicular to the skeleton segment), and also based on the area within the arc. These both require the setting of specific parameters, while the first-order retraction is parameter-free and, as has been seen, serves satisfactorily for the smoothing of many scanned images. In general they found the area approach to be more satisfactory.

Matuk et al. (2006) took the more general retraction approach in order to simplify not only a single curve (contour) but also a stack of contours, so that minor features such as small streams or

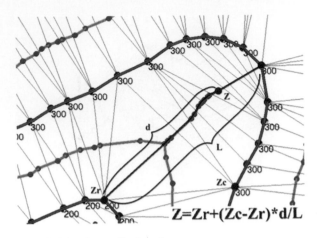

$$Z=Zr+(Zc-Zr)*d/L$$

Figure 158. Skeleton point heights from valley lengths.

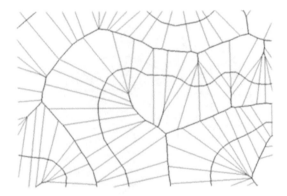

Figure 159. The 'Hidden Dual'.

ridges might be removed (feature-based simplification) without any lower contour being retracted so as to overlap the retracted portion of the upper contour. In principle this should be achievable, as the complete terrain surface has a 3D skeleton, although researchers have found it to be complex and difficult to work with. By working with individual contour-level slices through an approximation of the 3D contour a simplified method might achieve the same results.

4.1.5 Watersheds

As previously discussed, each VD/DT edge pair is tested to assign one edge to the crust, or the other to the skeleton. Nevertheless, the remaining edges still exist, even if they are not often used in the analysis process. An example is the Voronoi edge that penetrates the matching crust (or contour) segment: the 'Hidden Dual', as shown in Figure 159. The blue segments represent the contours, and the red ones represent the resulting skeleton, while the green ones show the penetrating Voronoi edges. These must, from the fundamental VD/DT relationships, pass through at right angles – indicating the direct down-slope direction of the surface.

The skeleton can also be applied to hydrography rather than to contours: if the path of a river is sketched in, then the watershed matches the skeleton sufficiently well for preliminary analysis. This follows from the constant slope assumption we used previously: if the rock types and erosion mechanisms are similar, then the valley walls will have similar slopes, and usually the watershed will be mid-way between the river valleys – although this will only be approximate: a major river will cut deeper than an adjacent minor one, and since the erosional forces are similar the

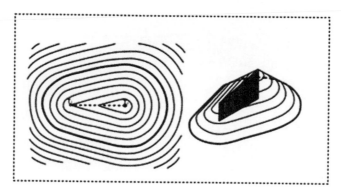

Figure 160.   Blum's Height Transform.

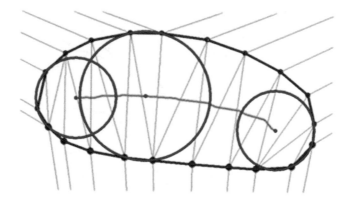

Figure 161.   The Height Transform from Voronoi skeleton circumcircles.

slopes will remain similar, but the height-of-land will be pushed further back. Thus each Voronoi cell indicates the surface area that will run directly into the river at its particular generating point.

As with the contour enrichment process, we now need to assign elevations to the skeleton/watershed points. As we have no reference contour elevations to work with we must use another approach, and go back to Blum's early work. One of Blum's additional observations was that points on the skeleton (which are vertices of the Voronoi diagram) may be assigned heights based on the radius of the empty circumcircles centred on these points. In (Figure 160) an ovoid generator has wave fronts that meet in the interior, to form the medial axis.

Blum's 'Height Transform' stated that the height of a medial axis point equalled the distance from its boundary: the highest points are the furthest away. Stated in Voronoi terms, the skeleton points are Voronoi vertices, with circumcircles that touch their triangle vertices (Figure 161).

Thus all skeleton points already have the appropriate circle radius, or distance from the boundary, and these radii are thus interpreted as heights. Figure 162 shows the simple interpolation model of a zero-elevation boundary and the interior elevated skeleton points.

Thus Blum's Height Transform allows us to estimate elevations for points on the watershed boundary – based on the fairly reasonable assumption that valley slopes are equal. The result is a TIN terrain model using river and watershed data points alone. Of course this is a rough approximation, based only on the river locations, but it often suffices to outline the geomorphic structure of the terrain. Figure 163 shows the sketched hydrography, and Figure 164 shows the watershed boundaries.

The complete Voronoi diagram is also of interest, as each cell outlines the area closest to each point on the river: as the non-skeleton boundaries are perpendicular to the river segments

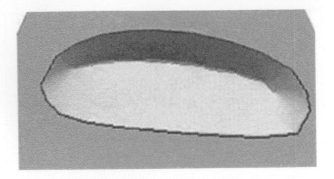

Figure 162.   Simple interpolation from the skeleton.

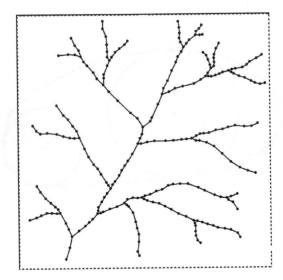

Figure 163.   Hydrography sketch.

Figure 164.   Rivers and watersheds.

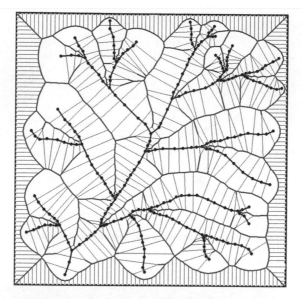

Figure 165.   Watersheds and Voronoi cells (external boundary added).

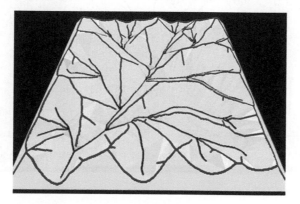

Figure 166.   Watersheds: 3D view.

(Figure 165), it is reasonable to postulate that rainfall falling within each cell runs directly down-hill from the watershed to the river itself – thus giving a preliminary model of total runoff. This can be accumulated along the length of the river itself, giving preliminary estimates of total (not time-dependent) runoff at any point in the river system.

While unrelated to precise interpolation techniques, this approach looks directly at the underlying hydrographic structure of the terrain, which is often difficult to extract from elevation data alone.

This may be considered in 3D if we initially take the river to be at a common base elevation, and estimate the heights of the points on the watersheds using Blum's method – and then generate a TIN using the Incremental Insertion Algorithm. Figure 166 shows the resulting 3D model.

Figure 167 shows the underlying triangulation, between the original hydrography (here assumed to be at zero elevation) and the elevated watershed.

The watershed is the boundary of the catchment area. The total catchment area is the sum of the areas of the individual Voronoi cells associated with each point along the various rivers. If we assume a fixed rainfall over the whole area, the water from individual Voronoi cells will flow to

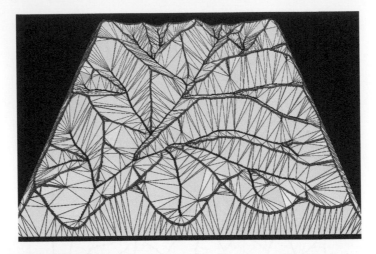

Figure 167.   Triangulated watersheds.

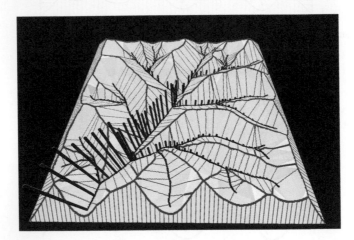

Figure 168.   Estimated total river flow.

the generating point on the river – from where it will flow downstream (along the crust/river) until it joins the next river, etc. and finally leaves the map area (Figure 168, Figure 169, Figure 170).

It should be noted that this is not a complete runoff model – there is no time element included. However, it could easily accommodate differing rainfall in different map regions, and it could easily serve as a quick-and-dirty evaluation of the rainfall, or flood, potential of a newly analysed area.

4.1.6   *Interpolating from contours: Sibson interpolation*

When these skeleton points with their elevation estimates are inserted into the TIN the results give a very reasonable assessment of the spatial structure of the terrain. Here are two views (Figure 171, Figure 172) of our contour data where we started with a triangulation model and used the useful parts of the skeleton as additional points added to the TIN.

If we want to improve on the look and surface continuity of our surface we should use Sibson interpolation (whose advantages were described previously) to interpolate a very fine grid – fine enough to observe the small-scale surface slope discontinuities.

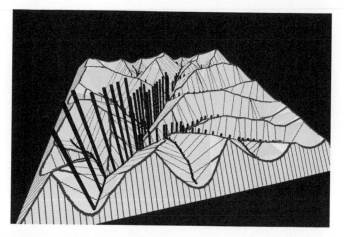

Figure 169.   Estimated total river flow – a second view.

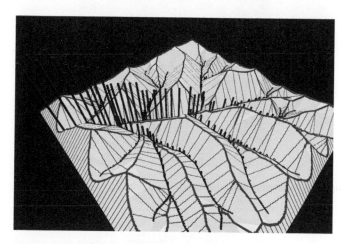

Figure 170.   Estimated total river flow – a third view.

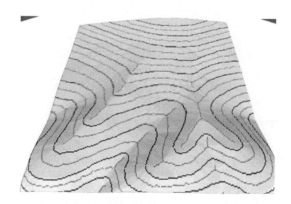

Figure 171.   Terrain with skeleton enrichment.

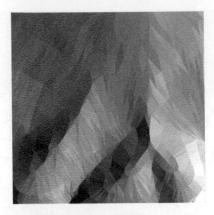

Figure 172.   Skeleton enrichment – illumination.

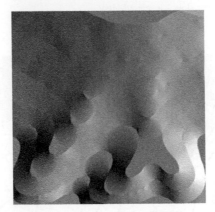

Figure 173.   Simple Sibson interpolation of contours – illumination.

Figure 174.   Sibson interpolation – with skeleton enrichment.

We process the contour data with simple Sibson interpolation and no slope estimates: the surface is smooth (C^1 continuous) except at the data points themselves – that is, at any data point along the contours, as can be clearly seen in Figure 173. Note that, as with the simple triangulation, there are 'flat' areas at ridge and valley lines, wherever all the Voronoi neighbours have the same height.

Figure 174 and Figure 175 show the result of performing Sibson interpolation on our enriched contour data set (with additional skeleton points included).

Figure 175.　Sibson interpolation – skeleton enrichment, illumination.

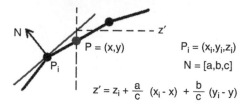

$$z' = z_i + \frac{a}{c}\,(x_i - x) + \frac{b}{c}\,(y_i - y)$$

Figure 176.　Interpolation with slopes.

Nevertheless, while the surface appears quite reasonable, there are still breaks in slope at the data points – in this case the contour lines and ridge and valley lines. Due to the area-stealing procedure, we are guaranteed to approach the correct data point elevation whichever direction we come from. Until we precisely arrive at the data point, however, we steal some of the cell areas from differing neighbouring points, and thus the surface may slope differently in each direction. We therefore need to add some slope information at each data point to improve the situation.

4.1.7　*Slopes: The forgotten need*

In real-world applications the accuracy of slopes is often more important than the precise elevation. This may seem surprising, but consider insolation (the amount of heating received from the sun), rainfall runoff down-slope, erosion, and vegetation coverage – quite apart from human activities such as cross-country trafficability and road construction. In all these cases results are not very sensitive to minor changes in elevation, but undulations in the terrain may be very significant.

However, undulations ('artefacts') often result from the interpolation process itself, as we have seen with some weighted-average methods. Sibson interpolation removes many of these problems, but needs to provide smooth surfaces at data points: it satisfactorily handles the first two components of a weighted-average interpolation – the choice of a set of neighbours, and the weighting function. The third component of weighted average interpolation is the function being interpolated: so far this has just been the elevation, considered as a horizontal plane through the point, but for better results a plane fitted to the data is a better choice, as it extends the 'best guess' of the surface behaviour near to the data point. Instead of using a weighted-average of the elevations of the neighbours, the average is taken of the values of the planar functions at the query location P (Figure 176). This process may be applied at local maxima and minima as well (Figure 177).

This allows slopes to be continuous on each side of the data point, converging on the planar function when the data point is reached, as shown below for the contour data. (Mathematically this

111

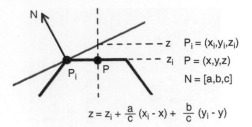

$$z = z_i + \frac{a}{c}(x_i - x) + \frac{b}{c}(y_i - y)$$

Figure 177.   Interpolation with slopes, at summit.

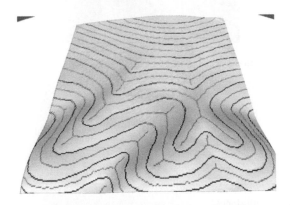

Figure 178.   Interpolation: Sibson, skeleton, slopes.

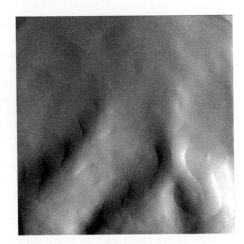

Figure 179.   Interpolation: Sibson, skeleton, slopes – illumination.

is not precisely correct – there are still small slope discrepancies, but so far, in practice, these may be ignored.)

(Note that this may produce modest extrema outside the range of the local data – especially at ridges or valleys, peaks or pits. This is often desirable, but in some cases – such as marine bathymetry – this may not be permitted as it could imply a greater sea depth than is actually available. In this case the horizontal plane is retained.)

Figure 178 and Figure 179 show the resulting surface when using Sibson interpolation plus additional skeleton points and slope estimates at each data point. The upper view makes it look

Figure 180.   Interpolation: TIN, skeleton, slopes.

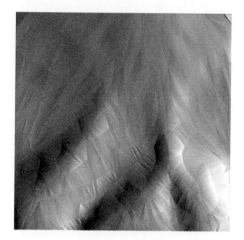

Figure 181.   Interpolation: TIN, skeleton, slopes – illumination.

completely smooth, but the lower shadow model shows minor remaining slope discontinuities, which may usually be ignored in practice.

In a few cases (e.g. geological strike and dip measurements) a planar function is already available along with the elevation data, but usually this must be estimated as a first step before the interpolation. Any simple regression through the data point is usually adequate, but a Voronoi-type approach, where the immediate neighbours of the data point are selected, may be appropriate. This provides a local planar surface, which does not usually pass precisely through the data points: this difference may be used as an error estimate – the 'jackknife' method, but in most cases the slope is used along with the original data point elevation.

However, Sibson interpolation is not the only approach that may be satisfactory: a simple planar interpolation in the triangle, when combined with the slope information, also gives surprisingly good results (Figure 180, Figure 181), and may well be the recommended method where algorithm simplicity is important.

Even the gravity model produces feasible slopes if the counting circle is carefully chosen – except close to the data points. Nevertheless, it is not usually to be recommended due to the difficulties in estimating appropriate counting circle sizes. Note the significant slope changes close to each data point, where the inverse-squared-distance weighting approaches infinity (Figure 182, Figure 183).

Figure 182. Interpolation: gravity model, skeleton, slopes.

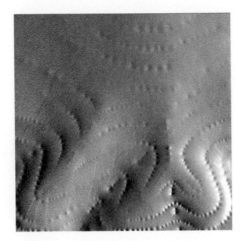

Figure 183. Interpolation: gravity model, skeleton, slopes – illumination.

Smoothness

Surface roughness (or its inverse, smoothness) may be estimated by using these local jackknife statistics around data points: this is an estimate of the surface roughness observed in the data, or the errors in the observations – they often cannot be distinguished without additional information. This is a plausible technique for smoothing irregularly-distributed data: replace the original observation by the new estimate and the result is a simplified surface – this process may be repeated several times if desired.

Plausibility

A final, catch-all quality for interpolated surfaces is 'plausibility'. As the name suggests, this can be highly subjective. A simple TIN structure, for example, may be reasonable in terms of surface continuity but implausible in terms of the breaks in slope at triangle edges – these are 'artefacts' of the construction method. Flat-topped hills are also implausible: we know that they should be rounded above the top contour level of our data – or else, in a different geomorphologic environment, they should have sharp peaks. We have seen a partial solution to this situation using the Voronoi skeleton. Another situation – again, particularly visible when using contour data – is an interpolation method that has slopes tending towards zero at data points: many weighted average techniques have this problem if slope functions at the data points are not included in the method.

This is clearly visible in perspective views, and has been found in national mapping agency DEMs where these were generated from old contour maps.

From all this we may make some generalizations about weighted-average terrain interpolation.

a) Avoid counting-circle methods, as usually these are in conflict with the weighting function: surface breaks occur when, as the query point moves, data points are rejected before their weighting is reduced to zero. Avoid gravity models and neighbour selection techniques which require user-specified parameters.
b) Neighbour selection should always use topological neighbours, such as the triangle or Sibson methods, where the weightings are restricted to the selected set of neighbours.
c) Wherever possible, assign slopes to data points.
d) For contour data the addition of skeleton points along ridges, valleys, pits, summits and passes is required to eliminate flat triangles.

FURTHER READING

A central concept in hydrology is the 'watershed' or 'catchment': the first refers more specifically to the 'height of land' and the second to the enclosed area – the first being the 'skeleton' of the Voronoi cells between river branches, and the second being the cells themselves, or their sum. Gold and Dakowicz (2005) used this model for easy initial hydrology modelling. This gave a feature-based view of the terrain.

Various approaches have been made over the years to extract good terrain surface models from contour lines. While this is of less importance in recent years due to satellite and LiDAR altimetry, there are still many situations where early contour maps are the only data source. Examples are Walters (1969) and Fuchs et al. (1977). Thibault and Gold (2000) used skeleton extraction to estimate ridge and valley forms where otherwise 'flat' triangles would lose the ridge and valley lines. Dakowicz and Gold (2003) summarized surface interpolation from contours, emphasizing the value of incorporating slope values at data points, while Matuk et al. (2006) simulated the 3D skeleton of the terrain in order to simplify the topographic features.

Another application that requires a 'reasonable' initial surface is for visibility computing on the terrain surface. Algorithms for this are De Floriani and Magillo (2003) and Sang et al. (2007).

4.1.8 *Runoff modelling*

As previously indicated under 'spatial models', a terrain surface may be thought of as a DT, a VD – or both together. A prime example of this is runoff modelling, whether of a raster grid or of irregular Voronoi cells. Finite Difference modelling, where a quantity of water moves from each cell to its downhill neighbours during each iteration, is an excellent example of this: it needs both a 'bucket' and a 'slope'. The bucket, of some known area, serves to contain the water; the slope indicates the downhill velocity. But: adjacent buckets have vertical boundaries! Thus, even for simple raster algorithms, there are primal and dual concepts involved – the square bucket, and a slope imagined to be between the centres of adjacent grid cells. The Voronoi runoff model is identical, except that the geometric structure of the Voronoi edges and the Delaunay edges must be explicitly preserved. Then, for each time step, the water transfer is calculated from the slope of the Delaunay edge, the area/volume of the Voronoi cell, and the common edge between an adjacent cell pair.

If we look at Finite Difference models we need to consider the 'buckets' holding water at some particular moment in time, the slope or height difference between adjacent buckets, and the lengths of their common boundaries. The most common approach uses a grid or raster surface model. The bucket size, the (four) neighbours and the length of the common boundary are constant for the whole map. Water will run downhill from higher cells to lower ones, crossing their common boundaries. Slopes are estimated by assuming that the centre of the cell represents the location

115

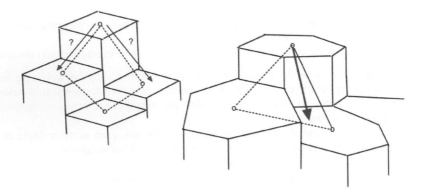

Figure 184.   'Block world' and 'slope world'.

of the whole cell: not an obvious idea, but better than having it fall vertically off the edge of the cell! Thus underlying the common 'four-neighbour' runoff models we have assumptions about the duality of graphs, with the Delaunay vertex representing the whole Voronoi cell. (Note that in this case the Delaunay graph is composed of squares, not triangles: these may be subdivided into triangles if convenient, e.g. for visual display.) The grid cells are the Voronoi cells of their central generating points, giving a 'block world' model of the surface. The Delaunay triangulation gives a TIN model: a 'slope world' (Figure 184).

For each time interval the flow of water from each higher cell to each lower cell is calculated, the transfer is made, and the process is repeated for the next iteration. Usually this is based on the four neighbours to each cell – which means that at a small scale flow is constrained to the NSEW directions, producing an implausible rectilinear flow pattern. This may be improved somewhat by adding the cells that just touch the central one at the corners, rather than the sides, but it is hard to justify this in Finite Difference modelling terms where flow passes through the boundaries of adjacent cells only. A particular problem with these grid approaches is the existence of small 'pits' where due to the limitations of the interpolation method some single cell may be slightly lower than all its neighbours – and all the water from the surrounding area flows into it.

An alternative is to use the Voronoi diagram of the original irregularly-spaced elevation data points, together with its Delaunay triangulation, to define the buckets, boundaries and slopes. The TIN model here has well defined triangles, and triangle edges to give slopes, whereas for the grid model the splitting of the original square Delaunay graph into triangles is not obvious. (Some research has been done on the flow of water down triangle faces, but the algorithms are rather complex.)

At some increase in computational expense the Voronoi method may be used to improve the results. As with the grid, the chosen interpolation method is used to make surface elevation estimates – but this time at random locations and Voronoi cells are constructed from these. Finite Difference modelling is then performed, where the flow between a pair of adjacent cells is proportional to the common boundary length and the slope between their generating points (based on the dual Delaunay triangulation). As the points are randomly distributed there is no reticulated pattern. The problem of 'pits' may be handled by allowing water to accumulate in each cell, as in nature, and then spill over. This is very similar to groundwater flow simulation.

As mentioned, runoff tests of this nature may be used to examine the terrain model for undesirable artefacts, and then improve the algorithm used. Figure 185 is a deliberately oversimplified surface model based on contours, with skeleton points added (Figure 186) and with randomly spaced interpolation points (Figure 187) – random points were generated, and inserted into the model if they were further than some desired spacing from their nearest neighbour. The Delaunay triangulation is shown below, along with the original contour data.

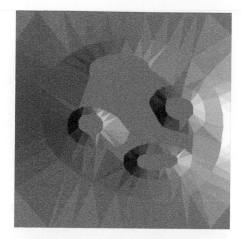

Figure 185.   Simple surface.

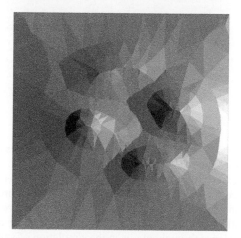

Figure 186.   Simple surface, enriched.

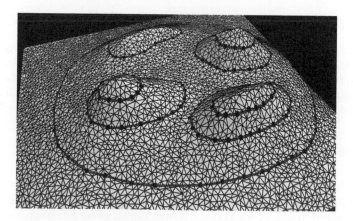

Figure 187.   Surface with random interpolated points.

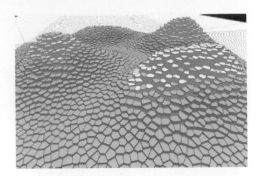

Figure 188.   Voronoi cells – early runoff.

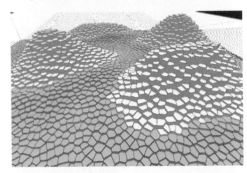

Figure 189.   Voronoi cells – late runoff.

Alternatively this may be viewed as the Voronoi cells, and the slopes of the triangle edges used to calculate the flow between pairs of Voronoi 'buckets'.

After several iterations, where the flow between any pair of cells was calculated as shown above, an initially uniform surface water depth after a rainfall appears as Figure 188.

And after more iterations the water has receded to the valleys (Figure 189).

One important application of a 'good' terrain model – one having reasonably defined ridges and valleys and plausible slopes throughout – is in the simulation of surface runoff (among other applications, such as insolation and vegetation patterns), so using a simple simulation like the one above can indicate the validity of our interpolation method.

FURTHER READING

Surface runoff and groundwater flow modelling using the Integrated Finite Difference Method may be based on flow between Voronoi cells, as performed manually by Narasimhan and Wither-spoon (1976) and implemented in a digital Voronoi model by Lardin (1999). Streit and Weismann (1996), and Sui and Maggio (1999) also discussed the integration of flow modelling with GIS. Dakowicz and Gold (2007) discussed various forms of flow modelling based on the VD/DT.

4.2   CLASS II: CLUSTER BOUNDARIES, OR 'FAT' VOID BOUNDARIES

4.2.1   *Clusters*

These are multi-point 'boundaries', where the boundary may be several points wide. They occur in two different contexts: firstly, as in the Cosmic Spatial Model, where galaxies/points have migrated towards 'slabs' at the boundaries of 'voids'. Here the emphasis is on the voids/polygons (polyhedra in 3D) that are being enclosed.

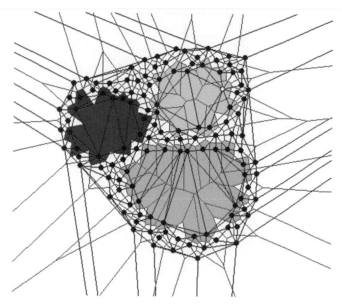

Figure 190.   Voids and clusters.

The second context is that of cluster analysis. While there are many different aspects and emphases, e.g. depending on whether the points have relevant attributes and whether the number of clusters is known in advance, the interesting aspect for us is the definition of the cluster boundaries, and perhaps the way they may be deduced from the forms of the Voronoi cells. After all, in the Cosmic Spatial Model we use the galaxies at the edges of the slabs to generate the portions of the gravitational force fields that define the polygons/polyhedra/voids. (We are simplifying a bit here: galaxies in the slab interior also contribute to the overall gravitational forces.)

For multiple-point boundaries (Figure 66) we are trying to establish the boundary points whose force-field zones define the voids/polygons of our map: we are labelling these points with the 'name' of the polygon. Consequently we have an external limit for the Voronoi cells of the polygon – the labelled skeleton separating the polygon cells from the 'rest'. This separates the two 'labelled crusts' – the connected sets of the contiguous boundary points whose fields form the polygons.

Note that in the Cosmic Model the cluster is formed from the set of connected boundary slabs, and has holes/voids within it (Figure 190). Nevertheless, each of these voids/polygons has boundaries formed by individual slabs, with the structure described above.

This works if we know the voids beforehand, and have labelled the points in advance. If we need to work this out from the geometry alone we need to look at Voronoi-based cluster boundary ideas (Figure 191).

Cluster analysis has produced a large number of methods and algorithms, which can't be discussed in detail here – partly because the definition of a cluster, and how much separation is necessary and what attribute information is available varies greatly between applications. However, for our purposes, one of the most powerful, yet simple, techniques in cluster analysis is the Minimum Spanning Tree. First the DT is constructed, and then the edges are pruned to produce the Euclidean MST, where each point is guaranteed to be connected to its closest neighbour. Then 'short' tree edges will connect points within a particular cluster, and 'long' edges will connect between clusters. The final objective is to remove 'long' edges between clusters, and leave the rest. However, what constitutes 'long' varies considerably between researchers and the type of data they are considering. Estevil-Castro and Lee (2000) put it well:

"There are four types of edges in Delaunay diagrams. There are edges within a cluster, between clusters, between cluster and noise and between noise points. The edges that should be removed at a certain level include all edges between cluster and noise, all edges between noise points, and some

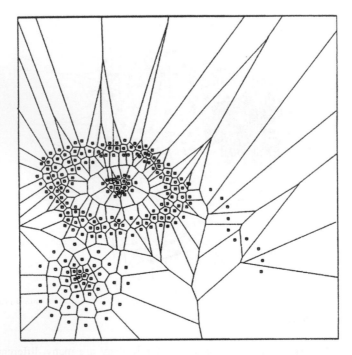

Figure 191.   Clusters and Voronoi cells.

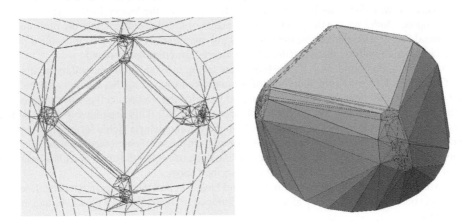

Figure 192.   DT of vector normals on the hemisphere.

edges between clusters. Noise points have a high mean adjacent edge length and a high number of edges; boundary points have a low number of edges and a high local mean; and cluster interior points have a low number of edges and a low local mean length.'

They based their work on the ideas of Zahn (1971): "The Euclidean MST may have k known clusters, or be found by deleting 'inconsistent' edges – edges whose weights are significantly larger than the average weight of the nearby edges in the tree."

The work of Tse (2008), described later, used these ideas to separate clusters of orientation data from the vector normals of the triangles forming models of roofs. These were projected onto the unit hemisphere and the DT constructed (Figure 192).

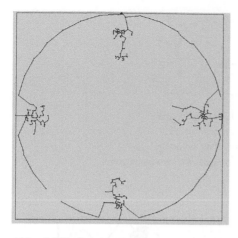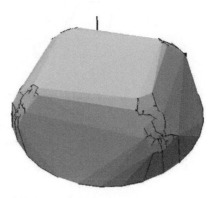

Figure 193.   MST of vector normals.

The Euclidean MST then indicated the clusters (Figure 193). As the separations on the hemisphere were measured in degrees of arc, a cut-off of 5 degrees was used to distinguish between points that were 'close' in angular distance and those that were 'far'. Each of the remaining clusters was then presumed to be a sample from a particular roof plane. Once the clusters have been identified the points may be labelled, and the labelled skeleton calculated as required.

FURTHER READING

Cluster analysis occurs in many disciplines, and it is difficult to describe all the methods. Gower and Ross (1969) and Rohlb (1973) described single linkage clustering from minimum spanning trees (MSTs) of arbitrary graphs. Zahn (1971) examined MST methods with particular reference to 2D point groupings, and Boots (1974) looked at the DT for point pattern analysis. Ahuja and Tuceryan (1989), and Tuceryan and Jain (1990) used the shapes of the Voronoi cells of points to determine if the point lay on the interior or edge of a perceptual cluster. Ng and Han (1994) used the VD to determine if a point was 'noise' or part of a cluster. Estivill-Castro and Lee (1999, 2000; Lee and Estivill-Castro 2002), in various papers, used boundary extraction to identify clusters, while Edla (2003) reviewed Voronoi-based clustering algorithms.

4.3   CLASS III: DOUBLE-POINT BOUNDARIES

4.3.1   *Labelled skeleton – map digitizing*

In Figure 67 the interior points of multi-point boundaries are removed, keeping only those points that were used to form the labelled crusts of polygon pairs – and thus also the skeleton between them.

Double-point boundaries have been found appropriate in some major map digitising projects, such as soil or forest maps and reports, where a large number of contiguous polygons need to be digitised and the underlying map is sufficiently complex that simple scanning will not suffice. In this case the cursor is 'rolled' round the interior of each polygon (perhaps with the aid of a small central dot) and a polygon label assigned to each of the points. This is repeated for all polygons on the map. When discarding all Voronoi boundaries between points with the same label – between adjacent points on the polygon interior – the boundaries between the polygons remain – in many

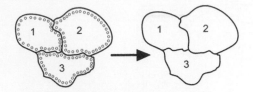

1: Select polygon label for interior points
2: Click points round polygon interiors
3: Make Voronoi diagram of all the points
4: Delete edges between points with the same label

Figure 194.   Rapid manual digitizing process.

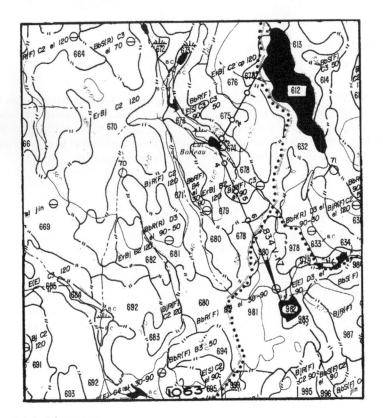

Figure 195.   Original forest map.

cases with enough precision for the application (Figure 194). (Efficient algorithms exist for this extraction and aggregation process.)

An interesting point is worth noting: because the original set of Voronoi boundaries is topologically connected, and because the same is true for contiguous aggregates of cells, the boundaries between the polygons are themselves guaranteed to be topologically complete, even taking account of digitising errors. This is particularly useful in the processing of scanned maps when using the labelled skeleton.

In this particular example it was necessary to extract the forest stand boundaries from a paper map where they were superimposed on other map data: it had been found too complex to attempt automated methods, and an experienced operator was needed (Figure 195). A compromise was

DIGITIZING FLOW

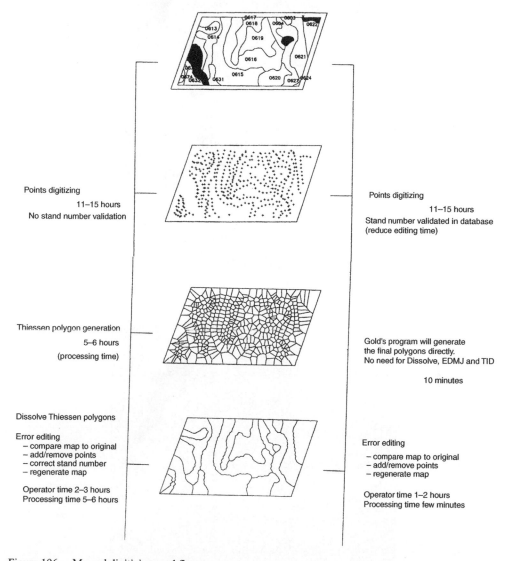

Points digitizing
11–15 hours
No stand number validation

Points digitizing
11–15 hours
Stand number validated in database
(reduce editing time)

Thiessen polygon generation
5–6 hours
(processing time)

Gold's program will generate
the final polygons directly.
No need for Dissolve, EDMJ and TID

10 minutes

Dissolve Thiessen polygons

Error editing
  – compare map to original
  – add/remove points
  – correct stand number
  – regenerate map

Operator time 2–3 hours
Processing time 5–6 hours

Error editing

  – compare map to original
  – add/remove points
  – regenerate map

Operator time 1–2 hours
Processing time few minutes

Figure 196.   Manual digitizing workflow.

needed: a 'rapid digitizing' technique was developed, whereby a small dot was placed at the centre of the cursor and rolled around the interior of each polygon – and the label changed for each polygon: the operator thus only had to untangle the polygon boundaries from the underlying map, and not worry about precisely connecting them without any topological errors.

The DT/VD of the point set was then constructed, all VD boundaries between points with identical labels eliminated, and the remainder plotted (Figure 196, Figure 197, Figure 198, Figure 199). The resulting boundaries were sufficiently close to the originals to serve the intended purpose of calculating the stand areas, and saved considerable operator and processing time.

123

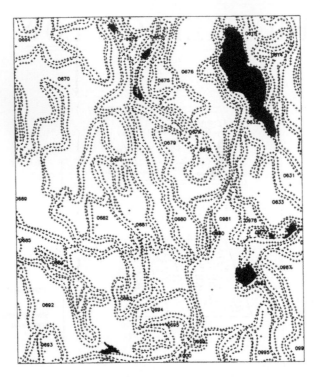

Figure 197.   Digitized points.

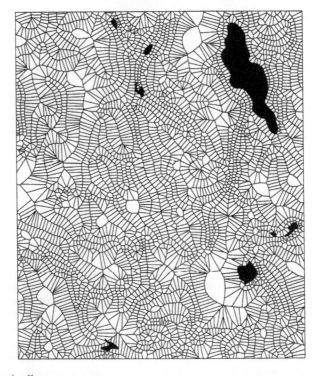

Figure 198.   Voronoi cells.

124

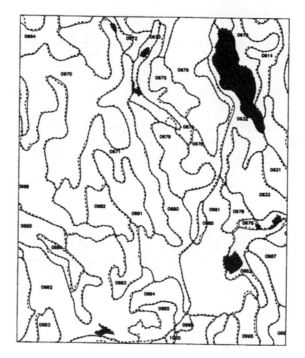

Figure 199. Resulting boundaries.

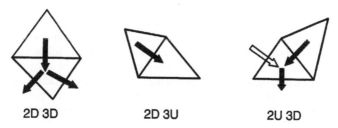

2D 3D   **2D 3U**   **2U 3D**

Figure 200. Triangle ordering.

It was necessary to traverse all the triangles only once in order to extract the dual Voronoi edges required. This required a sorting order for the triangles, so that each subsequent triangle was adjacent to one previously processed: a 'top to bottom' order – but points in 2D (or 3D) have no intrinsic ordering. Nevertheless, once they are triangulated it is possible to use the CCW test to identify each triangle's orientation with respect to some exterior point – the 'North Pole, NP' – and use this to define a partial ordering- a graph traversal – that visits each triangle – or each edge, or each vertex – once and once only. Figure 200 illustrates the method.

Each triangle edge has its vertices (in anticlockwise order) plus NP tested with CCW. If CCW is positive then the edge exterior faces downwards, and another triangle may be subsequently be added below. If not it is facing upwards and the branch of the traversal terminates. If we label the edge where the triangle was entered from above as 'Edge 1' then there are three possible triangle orientations, as shown. If it is '2D 3D' then there will be two subsequent downwards children: the triangle below 2D will be processed next, and 3D will be put on a stack (a list where the topmost, or most recent, is retrieved first) for later processing. If the triangle is '2D 3U' then no child triangle is processed, but if it is '2U 3D' then one child triangle below 3D is processed immediately. '2D 3U' and '2U 3D' are the same, but entered from different edges, so the single lower child triangle

125

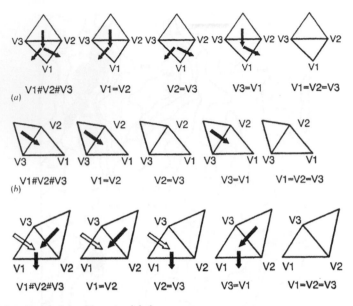

Figure 201.   Triangle ordering with vertex labels.

that was ignored in '2D 3U' is subsequently retrieved in '2U 3D'. This gives a traversal where new edges are added onto previous edges, without the need to flag where the algorithm has previously been, as is the case with most methods. In order to process boundary edges it is the DT dual (the Voronoi edge) that is retrieved.

This technique is also useful for the incremental drawing of contour lines, for example, after interpolation. It also applies in 3D, using oriented tetrahedra rather than oriented triangles.

When the boundary edge segments were output to a conventional GIS however, the processing time to add each segment to the growing boundary curve was found to be excessive, so an additional step was needed. For each of the three cases above there are five cases of similar-or-different vertex labels, giving 15 cases in all (Figure 201). These indicate whether portions of a boundary are started, continue or join, and the appropriate lists can be updated immediately.

4.3.2   Scanned map processing

The double-point boundary, where each row of points is associated with one polygon, occurs in various applications. Here is another legacy application, where polygon boundaries need to be transferred from paper maps – but the map complexity is low enough that automated methods may be used.

First the map is scanned to a fairly high resolution, and then an edge-detection filter – common in many remote sensing packages – is used to find the high contrast edges between 'black' (ink) and 'white' (paper). These raster cells may then need to be thinned, again using a standard filter, before the VD/DT is constructed. The crust and skeleton are then extracted, for a small polygon set (Figure 202).

The skeletons are then smoothed (Figure 203), as described previously, using the skeleton-retraction method, which guarantees the preservation of the topology. The skeletons of the 'white' polygon interiors give some information about the polygon shape – one might here imagine the roof of a house – but the particularly valuable part is the skeleton of the interiors of the 'black' line-work. If the line-work was thick enough to separate the two boundaries then the interior skeleton is guaranteed to be topologically complete, as was mentioned for labelled skeletons.

The results are particularly interesting for the digital conversions of plans and cadastral maps as, in Figure 204, the first image gives the raw crust and skeleton, and the second the result after smoothing by skeleton retraction (Figure 205).

Figure 202.   Crusts and skeletons for a small polygon map.

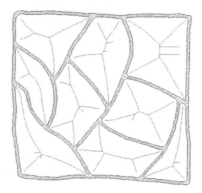

Figure 203.   Polygon map, smoothed.

Figure 204.   Scanned cadastral map.

Several points should be noted.

a) The skeleton of the line-work is topologically complete, even at junctions (two-point boundaries always have a separating skeleton).

b) The skeletons representing the interiors of houses indicate approximate possible roof plans, based on assumptions of constant slope (as was described previously for watersheds).

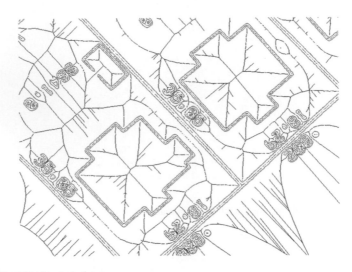

Figure 205. Smoothed cadastral map.

c) The skeletons around the buildings show the relationships between the buildings and the property boundaries. (Only the VD is shown; the DT segments for the skeleton edges will give the linkages between the building and the property boundaries.)
d) The characters of the drafted labels may be recognised with the aid of their skeletons, as described earlier.
e) The characters may be assembled into 'words' using the adjacency of their exterior skeletons.
f) These labels may be associated with their appropriate polygons or boundary segments using the adjacency of their exterior skeletons.

Quite a lot of contextual information may be derived from the VD/DT and a couple of minor algorithms. The dual is the context!

FURTHER READING

Gold et al. (1996) described and implemented a rapid polygon map digitizing system that resolved a significant industry bottleneck by using the labelled skeleton, and Gold (1997) used an edge-detection procedure, together with the crust and skeleton approach, to scan cadastral maps automatically.

4.3.3   Hierarchical VD – Spatial indexing

When the number of points gets very large the simple algorithms discussed so far start to exceed computer main memory, and out-of-core techniques need to be considered.

Traditionally we have a coarse grid whose cells contain finer details, so we look for the grid cell bracketing the coordinates of the desired point. In a variant, Quad-trees, the map is subdivided into hierarchical rectangles as points are added. However, these are all based on a subdivision of space into coordinate-based blocks, which can have difficulty handling very anisotropically distributed data. The VD/DT, on the other hand, adapts itself to the data density, but is not hierarchical: it is conventional knowledge that one of the disadvantages of the VD is that there is no hierarchical structure available, so that rapid search of very large data sets is slow.

Nevertheless, various kinds of searching may be performed, such as the simple Walk through a triangulation, as described earlier. (Most of the time this may be used for any triangulation, but cases exist where this does not work. However, it always works for the Delaunay triangulation.)

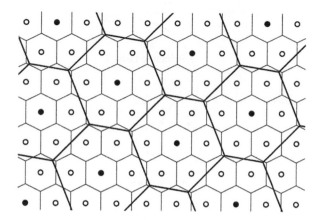

Figure 206.   Christaller Voronoi hierarchy.

Another is the Visibility Ordering traversal which guarantees that triangles (or edges, or vertices) may be traversed in a front-to-back order from some specified viewpoint, as previously described. This is useful for a variety of applications, such as drawing a TIN model in perspective view, where we wish the front triangles to be drawn before the rear ones, which would be clipped by those in front (or, alternatively, when using the painter's algorithm in a raster system, to draw the rear ones before they are overdrawn by closer ones). It is the basic algorithm in the edge removal process previously described for forest maps.

Nevertheless, the maps to be processed are still limited by the size of the computer's memory. Inserting a block of points within a lower-level Voronoi cell will not produce smaller Voronoi cells that fit within the larger one, and the same is true for Delaunay triangles: what to do?

The Voronoi diagram (VD) partitions space into cells whose interior points are closest to the generating point. The Delaunay triangulation (DT) is the dual of this – its edges connect generators that have adjacent cells. In the Simple Incremental Algorithm we can walk through by using the CCW test and following the VD or DT edges, as described earlier. In terms of algorithm efficiency, for n generators the insertion/switching stage takes constant time, as the expected (average) number of adjacent Voronoi cells is six, limiting the usual number of switches. However, the search through the triangular mesh could take n^2 tests, or $n^{1.5}$ for the 'Walk' algorithm described earlier – that is, for searching for all n points. This is not very efficient – an ideal search through a tree-type index would be $n(\log(n))$ – but the VD cannot be made directly into a hierarchy – lower level cells will not fit exactly into higher level cells.

A solution can be found based on Christaller's (1933) work, where he examined town and village distributions and their respective hinterlands: he did not attempt to insert lower level village zones of influence within the higher-level zones for towns. He inserted the centre of the zone – the village location – instead: clearly points can easily be inserted within a town's higher-level index zone, and yet lower level points can be located within a village zone: his administrative optimization model (Figure 206).

This implies that a cell is within a higher level index cell if its generator is inside, and thus we can 'walk' from cell to cell as described before, then drop down to the next level and continue. Each index generator has a pointer down to some child generator in the index cell – this gives a logarithmic search time, as desired (Figure 207). The Voronoi problem can thus be solved by using high-level Voronoi cells as index cells for the next level of points, letting their cells be secondary indexes for the third level points, and so on (Figure 208). In searching for a particular location a walk is performed at the top level until the enclosing cell is found, along with a pointer to some point in the next level down (if any exist), repeating the walk in this second level, and repeating the process until a closest point is found at the base of this search tree. This speeds up the theoretical time for the walk (and hence the whole incremental algorithm) to $n(\log(n))$, and any desired index

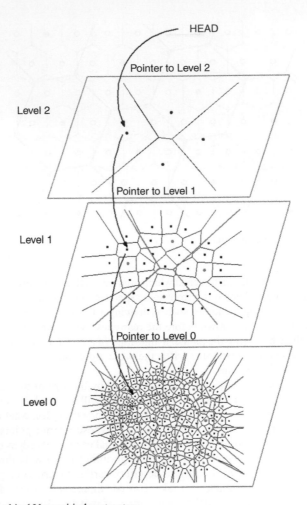

Figure 207.   Hierarchical Voronoi index structure.

level may form the basis for separating the map 'sheets' and storing them in separate portions of the hard drive.

Index generators

Any desired method may be used to get generators at the index level: they may be sampled from the lower level generators; they may be on a grid; they may be random. (Special care is needed if an index cell is nearly empty.) In our implementation we used the Quad-Edge data structure, as it stores both the primal and the dual links, and we walk along the edges to locate our desired points. As the structure is based on edges, we can cross from one disc page to the next by following duplicate edges – those which were split by the page boundary (the previous index level's cell edge), and which thus have to be stored on both of the adjacent pages (Figure 209).

If our data will not fit in memory we must split it into disc pages: each index cell contains one page. When searching we walk along the top index level and then drop down to the page (index level) below. Sometimes we must cross from one page to the next. For this we keep duplicates of the edges that cross the boundary, and its end point on the other side. This tells us which page to go to, to find the matching duplicate edge. Searching then continues. The same approach may be used for parallel processing: this method puts spatially adjacent points on the same disc page, which is not usually feasible.

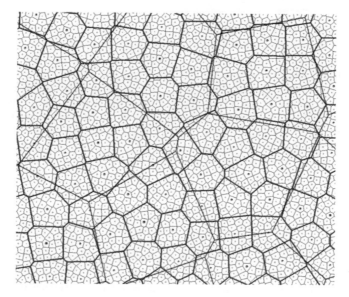

Figure 208.   Three-level hierarchy.

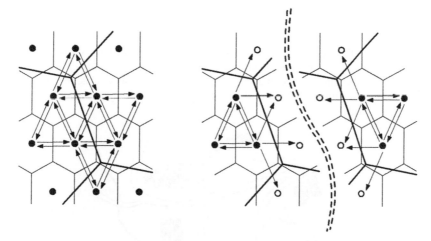

Figure 209.   Hierarchical paging system.

FURTHER READING

Spatial ordering or indexing of points and triangulations is a fundamental requirement of efficient terrain algorithms. The fundamental operation is the ability to 'walk' from one triangle to the next adjacent triangle (or, equivalently, to the next edge or vertex). To our knowledge, this was first described in Gold et al. (1977), and then elaborated on in Devillers et al. (2002). Gold and Maydell (1978) showed any Delaunay triangulation may be ordered as a binary tree when considered from any fixed viewpoint. Gold and Cormack (1987) used the binary tree order to process triangles (and contour segments) in a top-to-bottom order, and De Floriani (1989) and De Floriani et al. (1991) described other algorithms for Delaunay triangle ordering. De Berg et al. (1996, 1997) formalized the traversal of any planar subdivision.

Gold et al. (1996) described an efficient manual digitizing method for polygon sets, based on the binary tree order, and Gold (1997) automated this for closed polygon sets for 'clean' maps. Gold (1999) and Gold and Snoeyink (2001) – discussed above – extended this to general line sets by using the crust and skeleton rather than labelled points.

Once a 'flat' navigation system is available, faster walking from one location to another may be made more efficient with a hierarchical indexing system. Dutton (1996) described a hierarchical mesh indexing system with different levels of nested triangles, and van Oosterom (1999) surveyed the various GIS spatial access methods, as did Snoeyink (1997). Finally, Gold and Angel (2006) produced a true hierarchical Voronoi index based on the ideas of Christaller (1933), which permits order n(log n) searching.

4.4   CLASS IV: SOLID LINE SEGMENTS – HALF-LINE PAIRS

4.4.1   *Complex objects*

So far we have been concerned only with point objects – but the real world (in 2D) is composed of objects with area. Indeed, the Voronoi diagram we have been examining has difficulties with precise points: while necessary for calculations, the detection of collisions is made more feasible if we consider the points to be circles with a small radius – the 'tolerances' used.

But for real-world objects we must go further. Indeed, we have already done so. In the discussion of crusts and skeletons we saw that we could define, thanks to Amenta, a minimum spacing for points around a curve that would preserve the boundary. This was also used in the Free-Lagrange tidal modelling exercise.

But this is still less than ideal: points only approximate a curve, or even a straight line. We would like our Voronoi/Delaunay generators themselves to be more complex – starting with straight lines.

Figure 210 shows the partitioning of a small village into component parts, the estimation of the Voronoi cells for each (Figure 211), and the resulting analysis of adjacency – which houses are adjacent to which road, for example.

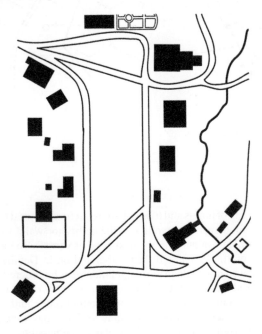

Figure 210.   Sketch of a small village.

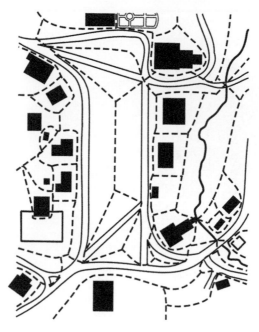

Figure 211.   Voronoi cells of the village.

Here each object – house or road segment – has a (hand drawn) Voronoi cell. And each road segment has two partial cells, one for each side of the road, exactly as for our previous examples: one could presume an additional skeleton segment along the road centre. But the objects are constructed of fully impermeable entities – the line segments – unlike our double-point boundaries, which are perhaps semi-permeable: points greater than the local tolerance could not penetrate them, but perhaps points with a much smaller tolerance could do so – line segments can perhaps be visualised as double-point boundaries – where the point density is infinite.

So we have two sides to each boundary, and each side has an associated void and also an adjacent side of another void. In addition, there is a Voronoi separator – a boundary to the half-line's cell – between each pair of half-lines, producing a form of 'sandwich'. Figure 67 shows the double-point boundary and the boundary separating the two polygons.

Figure 212 shows the polygon separator as a set of half-line sandwiches, with the polygon boundary running down the middle.

4.4.2   *The constrained DT and the line segment VD*

It is of course frequently necessary in GIS to represent objects with predefined boundaries – often linear – such as buildings or roads. A common approach is to extend the idea of the TIN so that some triangle edges conform to the building boundaries – the 'constrained Delaunay triangulation' (CDT). Various algorithms have been proposed: we will modify our Moving Point VD/DT algorithm to achieve this. While not necessarily the most efficient, it is consistent with the spatial model we have been discussing.

It is, in fact, a very simple extension: as the moving point MP moves, split off from some starting point, triangle edges are switched to maintain the VD/DT properties. If we want to follow and preserve some predefined edge we forbid the 'switching out' of that edge, always preserving its connection between the starting point and MP. This may be part of the drawing of an ongoing object boundary, or it may join the active line segment to some pre-existing point. While the 'split' and 'merge' operations are tricky, requiring tolerance checks and the detection of other edges connecting

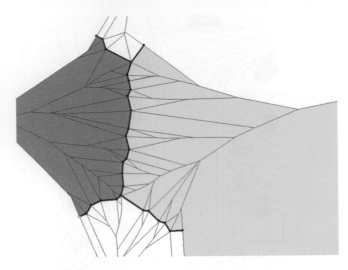

Figure 212.   Solid-line polygon boundaries.

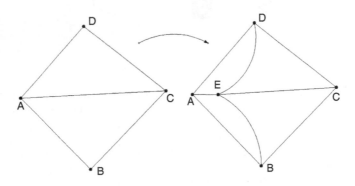

Figure 213.   Splitting off a new moving point.

to each vertex, the actual line generation operation is straightforward – and it is reversible, allowing the edge to be deleted.

New point E is split from A, and 'Zero-length' edge AE and two 'zero-area' triangles ABE and AED are created (Figure 213).

Two simple cases are shown (Figure 214, Figure 215) – the true DT edges shown in black and the constrained edges in yellow. The dual VD is also shown in Figure 214: note that the VD edge dual to a constrained edge is invalid – a necessary switch was prohibited, leaving overlapping regions. Thus the VD cannot be used with the CDT.

Figure 215 shows a more typical edge, perhaps attempting to represent the bed of a river valley within a TIN.

Because the edge construction operation is reversible, the intersection of constrained edges is manageable (Figure 216): if an active growing line meets a previous constrained edge – which may not be switched – this second edge may be 'un-drawn', and the various portions redrawn to the intersection point.

A fairly typical CDT of a set of buildings is shown in Figure 217. Note one important point: triangle edges serve two distinct functions – as building edges (green) and as the usual TIN adjacency markers (black). This breaks a very desirable property of a spatial data structure: the separation of the 'geometry' (coordinates) from the 'topology' (connectivity and adjacency relationships).

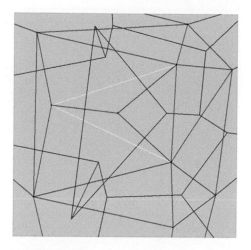

Figure 214.　Constrained DT edges and VD.

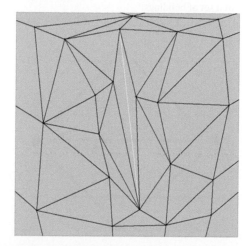

Figure 215.　Constrained DT edge.

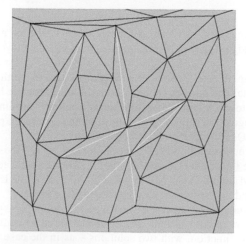

Figure 216.　Intersecting constrained edges.

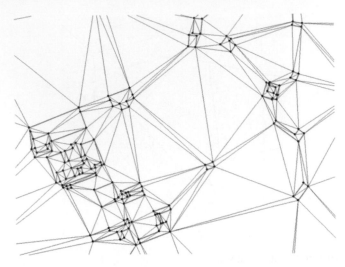

Figure 217.   Constrained DT of a set of buildings.

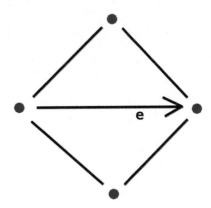

Figure 218.   A simple DT edge.

When a constrained edge is constructed, a triangle edge is labelled and serves both as a map object and as a linking edge (Figure 218): the points are in one data layer and the triangles/edges in another.

If we wish to keep map objects in one layer and linking edges in another we need to distinguish between an edge (triangle) and an edge (object). Since an impermeable boundary is two-sided, as just discussed, we need to work with half-edge objects, each of them linked to each other and to vertices by the dual triangulation (Figure 219). (Note that the arrows are purely indicative: triangle edges are formed using Quad-Edges, as described earlier.)

This allows the construction of the VD of points + line segments – the LSVD. Figure 220 shows a constrained edge: note that the VD is invalid. Figure 221 shows the equivalent LSVD: the line contains the accumulated neighbours of MP. Figure 222 shows a simple LSVD – note the existence of unconnected nodes and line segments, and intersecting line objects.

Note several differences. The boundary between a pair of points, or a pair of line segments, is a straight line – but between a point and a line is a parabola (the curve equidistant between the two). A parabola is a quadratic function, with two solutions – as in the case of its intersection with a line. The bottom of the figure – a point sandwiched between two lines – has two solutions – two

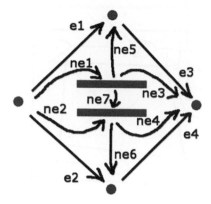

Figure 219.   Half-line structure.

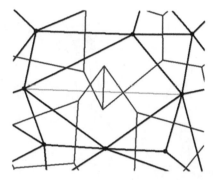

Figure 220.   A constrained DT edge.

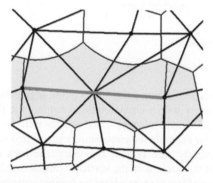

Figure 221.   A line-segment VD.

circumcircles for two anticlockwise triangles: (top edge – bottom edge – point) and (top edge – point – bottom edge). These must be distinguished in the arithmetic. In addition, a half-line and its endpoint are two distinct map objects, with a Voronoi boundary perpendicular to the segment, at its end. The circumcircle itself is calculated differently – it is 'tangent' to any combination of three line segments or vertices.

As discussed earlier, the kinetic point VD may be robustly maintained without great difficulty, and individual independent line segments created. The devil, however, is in the details. Two particular issues stand out. Firstly, a Voronoi vertex is equidistant from three generators – which may be three

137

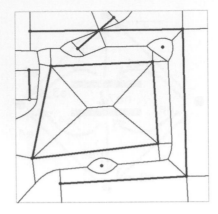

Figure 222.   An LSVD diagram.

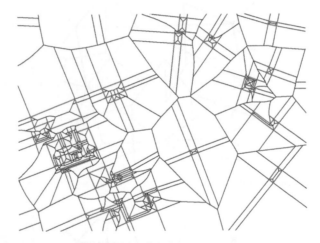

Figure 223.   LSVD of the set of buildings.

points, two points and a line segment, one point and two line segments, or three line segments. The intersection of the pen trajectory with a circumcircle is a quadratic function, with two solutions: where the trajectory enters, and where it leaves. This requires well defined and robust calculations, which can be problematic – for example the circumcircle associated with two joined line segments and a point, which has precisely zero radius! Secondly, the tests require a knowledge of the orientation of the tangents of the circle with the vertices – and so our best solution was an iterative one, where an initial selection of test tangent points (in correct anticlockwise order) is refined until the final circumcentre is within a small tolerance of the previous one. This has proved successful for our purposes. Development of a robust LSVD has been a time-consuming activity over several years, first by Yang and Gold, and then by Dakowicz and Gold. While effective, it has not had the extensive testing of the work of Held (2001) which, however, is a static algorithm, requiring reconstruction after each modification.

Figure 223 shows our set of buildings with the LSVD defining their boundaries rather than the Constrained DT as shown previously.

In Figure 224 the buildings' VDs are coloured, showing the adjacency relations between them. The buildings' skeleton boundaries, being equidistant to the buildings, provide a means of navigating between them, and a method for determining the largest vehicle possible along each path – a

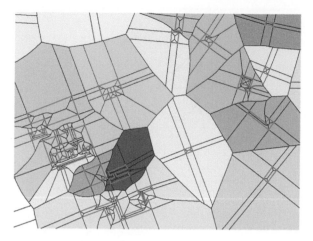

Figure 224.   LSVD of buildings, showing skeletons.

fire truck, for example. (Remember that points along the skeleton are centres of circles that touch the adjacent objects.)

The Iterative InCircle test

Our two-dimensional algorithm for moving points needs to be modified to handle line segments. Our triangles may now have either points or line segments at their vertices, instead of just points. As a result, a more elaborate circumcircle calculation is needed: instead of a circle touching the three points forming the triangle vertices, we must calculate a circle that is tangent to any line-segment vertices as well.

The work of Anton et al. (1998) shows how this may be done in an iterative fashion – indeed, it may be used with any objects at the vertices of the triangle. In the case of line segments, it is also necessary to know the side of the line segment that is connected to the triangle: this may be done with the half-line data structure.

a) Trim and/or extend segments (this has been modified from the original work)
b) Initial centre at mid-points of line segments (and/or at free or end points)
c) Calculate circumcircle using InCircle
d) Project centre onto line segments
e) Repeat iteratively to convergence

This guarantees the centre is on the correct side of the line, as InCircle rejects the test if the three vertices are in clockwise order. (This is how the two different solutions for the parabola/line intersections are distinguished, as mentioned previously.) Projecting the centre onto the (half) line segments also leaves the new vertices in the same (correct) order. Three examples are given: the initial large circle shrinks with each iteration (Figure 225, Figure 226, Figure 227). Because we are working with line segments, not infinite lines, the new vertices (and the initial ones) must be carefully trimmed so that the projections remain within the segments. Once a valid circumcircle is found the VD construction continues as before, with the dual DT preserving the spatial relationships. Other workers have developed different methods, but the iterative method, if somewhat slow, preserves triangle orientation in a guaranteed fashion by using the well-defined CCW and InCircle predicates.

The result of these extra steps is a spatial model that can maintain the VD for points and line segments, permitting the construction of a wide variety of map objects on the 'Entity Layer' while isolating them from the topological 'Adjacency Layer.' This property may be used to store map objects (and their attributes) in a database if desired, without breaking the connectivity of the Adjacency Layer (the triangulation).

139

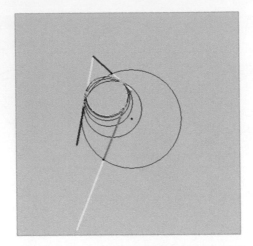

Figure 225. InCircle test 1.

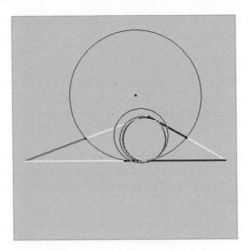

Figure 226. InCircle test 2.

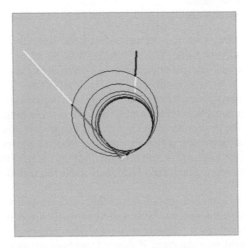

Figure 227. InCircle test 3.

Following Einstein's earlier comment, the map object, plus its Voronoi cell, or spatial extent, defines the local context of the object – as preserved in the dual triangulation.

FURTHER READING

Chew (1987, 1989) defined the constrained DT (CDT) and produced a divide-and-conquer algorithm, of optimal O(n log n) efficiency, for constructing the DT where some 'constrained' edges were pre-specified. De Floriani and Puppo (1992) described an algorithm based on the stepwise refinement of an existing triangulation by the incremental insertion of additional points and line segments. This is still commonly used for the modification of a terrain model (a TIN) to add ridge and valley lines.

However, it has several problems. The Voronoi diagram is not valid where constrained edges are inserted, and by having a triangle edge represent a map feature such as a building edge we are confusing the distinction between map objects and the underlying topology preserved in the triangulation.

Wang and Schubert (1987) defined a duality between the CDT and a form of the VD, and Gold (1990) described initial algorithms for the construction of the line-segment VD (LSVD). Within computational geometry, various authors in the late 1970s, 1980s and 1990s described the properties and algorithms for the LSVD – for independent line segments, for connected polygons or sometimes for intersecting segments: for example Burnikel et al. (1994) and Imai and Sugihara (1994). In most cases the end points are inserted first and then the line segment interiors. Gold et al. (1995) and Anton et al. (1998) described a completely incremental algorithm. Gold and Condal (1995) – described above – showed some early LSVD results. Held (2001) described what is probably the most robust and reliable LSVD algorithm to date, along with Karavelas (2004) who used the CGAL computational geometry system. Dakowicz and Gold (2006) and Gold and Dakowicz (2006) described the full LSVD incremental algorithm, along with various cartographic examples. The problem with all of these methods is the extreme complexity of the degenerate cases which must be handled when line segments approach, meet or intersect – in all cases many years of work was required, even though the normal use of the system, and the analysis of the results, is largely intuitive.

Research is still ongoing: https://github.com/aewallin/openvoronoi summarizes recent developments.

The 'Kinetic VoronoiMagic' program and manual at: http://www.voronoi.com/ is good for experimenting with the concepts, but its robustness has not been fully tested. Gold (2010) describes the latest work.

In an operational context Martin Held's VRONI package http://www.cosy.sbg.ac.at/~held/projects/vroni/vroni.html is recommended.

4.4.3 Basic queries with the line segment VD

The following figure illustrates some of the useful properties of a space-filling map coverage, rather than just a line-work spatial model. A simple polygon, and its Voronoi cells (and half-cells) is shown. A fundamental GIS question is: 'What is here?' If we only have line-work, or a discrete set of objects, this is usually unanswerable – the answer is: 'Nothing!' unless the query point falls (implausibly) exactly on a line. For a space-filling model – a 'field' model – there is always an answer: the cell within which it falls. The Voronoi cell is, by definition, the region closest to a particular map object, so a simple 'walk' through the mesh will arrive at the region for the 'closest' object to the query point. 'What is here?' becomes 'What is closest to here?' – which is often precisely what is needed.

In Figure 228 point P is in the interior half-cell of a segment of the polygon boundary, hence identifying the polygon as the target. Point Q falls within the (interior) region associated with a boundary vertex, and gives the same result. Points S and R are similar, but fall on the exterior

141

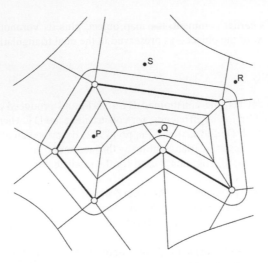

Figure 228.   Queries on an LSVD.

(half) cells, giving a response that 'The closest object is – the polygon.' Thus basic queries, and point-in-polygon queries, are directly resolved once the VD is constructed.

A further basic query resolves itself directly from the VD: "Find all the objects within (50 m) of the lake, or boundary," or "Is point P within (50 m) of the lake?" (This presupposes that it is not closer to some equivalent water body.) The VD divides the map into the 'closest' regions, so any point in the cell is closest to its generator – point or line segment. So, rather than constructing 'sausages' around boundary segments and then merging them, we simply find the 'closest' cell containing the point, and calculate if its generator is within the desired distance. (None of them are, in this example.) If the 'buffer zones' need to be explicitly represented then a (50 m) line or circular arc is drawn within each appropriate Voronoi cell.

4.4.4   *Dynamic map partitioning*

One of the significant issues with a GIS, and especially maps, is how to partition them without breaking individual elements – particularly the linear elements, which may vary from very small to very large. (We have already looked at the hierarchical partitioning of point data sets in the Voronoi diagram.)

Ideally the map should be 'cut' around well-defined entities, such as clusters of polygons, so that these elements may be pasted back in when necessary. But in order to do this we need to know how the particular boundaries of the entity match up with the 'hole'. One approach to this was based on the LSVD, where entities bounded by line segments could be removed, and the pointers (that is, the Delaunay edges that connect across the cut) were reorganised to point to their adjacent neighbours within the boundary. As the LSVD operates by splitting each line into half-lines, with the triangulation connecting each half, this may be readily achieved, and when the patch needs to be sewn back into the hole it is only necessary to save a single matching pointer from the boundary and from the hole within the tree structure. Thereafter a zipping process may be followed to reconnect the two portions. The resulting tree structure may then be saved in an appropriate database, with the root of the tree being the boundary container within which smaller, more detailed portions may be re-inserted. This work was described in Yang and Gold (1996).

Figure 229 shows a line-segment based map with five embedded sub-maps.

In Figure 230 the sub-maps have been cut out and the interior boundaries sewn together.

And here three of the sub-maps are shown (Figure 231, Figure 232, Figure 233). Topological connectedness of the boundaries is preserved by linking the half-edge pointers together around the exterior – but note that navigation is not possible outside the sub-map.

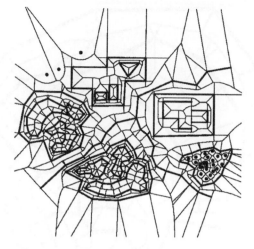

Figure 229.   A line-segment map.

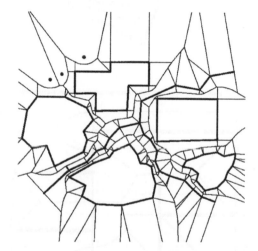

Figure 230.   Sub-maps extracted.

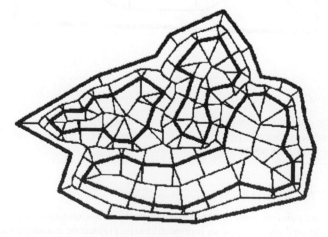

Figure 231.   Sub-map 1.

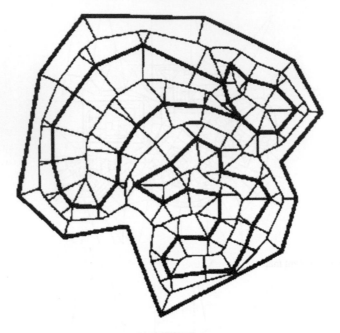

Figure 232.   Sub-map 2.

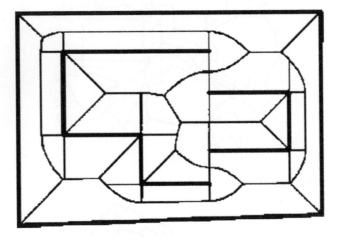

Figure 233.   Sub-map 3.

FURTHER READING

If the LSVD map contains enclosed regions, then a simple map partitioning method, redirecting the LSVD pointers, suffices to provide a useful map partitioning mechanism. Yang and Gold (1995, 1996) describe this approach.

4.4.5   *The kinetic line segment model and map updating*

Once we have kinetic point movement we have dynamic object maintenance, whereby line segments may be inserted, deleted and split on an individual basis. (Dynamic maintenance of points alone can be managed with only the insert/delete functions of the dynamic algorithm.)

A final interesting issue from this work is the perception that all these operations are based on a series of commands – a function name plus the object pointer, together with several parameters, for example the starting and ending point of the trajectory. All of these commands are reversible – they are their own inverses – and so a map may be considered as being constructed from a series of commands – a log file. A map may have a time stamp for each modification, and the map for any desired time may be produced, not as a series of snapshots but as an initial or intermediate map state plus the sequence of commands until the desired display time is reached.

This means that we can maintain objects by using a log file, which passes the appropriate parameters to the kinetic algorithm itself. This in turn means that we can apply a time stamp to each operation, making our map creation simulate the change in the situation on the ground. In addition, as each operation has an inverse (undoing a previously drawn line, for example) a partially or fully completed map may be rolled back to some previous time state.

While simple roll back/roll forward is straightforward, logical complications can ensue. If a roll back is followed by modifications followed by an attempt to roll forward again, then portions of the map that were previously used as starting or ending points to a line construction may no longer be available. This 'time-travel' conundrum has sometimes been referred to as the 'What happens if I go back in time and kill my grandmother?' problem. See Mioc et al. (1999, 2012).

This problem is particularly acute in cadastral mapping, because property maps not only require regular updating, but they also often need emendation due to surveying or recording errors, or legal decisions that affect the legal ownership of a parcel (including its merging or subdivision) dating from some time in the past. As with all databases, 'real time' and 'database time' are not always synchronised – the report of a property boundary change may be delayed, or forgotten, and needs to be inserted after – sometimes long after – subsequent adjacent updates. (It should also be remembered that the modification of a map object affects the Voronoi cells/polygons and adjacency relationships/triangulation of nearby objects, and in some cases, as for the insertion of a long road segment, the changes may be quite extensive.)

Finally, legal requirements often require the retrieval of the ownership status of one or several parcels many years in the past. It is therefore necessary to develop a mathematical framework for what modifications are acceptable, and how they are to be implemented.

However, the addition/deletion of a point or line are reversible (allowing reversion to a previous map state) and each of these are composed of a known set of atomic operations on the data structure. The possible operations are: Split (splitting a point into two); Merge (merging a point with its nearest neighbour); Switch (flipping the diagonal of a pair of Delaunay triangles); Link (adding a line segment between a pair of points after a split); and Unlink (removing the connecting line segment). These are the only operations that modify the data structure – a 'Move' sub-command executes a series of switches. Here 'Map Commands' (move a point; add/delete a point; add/delete a line); join/unjoin two points; join/unjoin a point and a line; and join/unjoin two lines) are composed from these and shown to be a unique set of atomic operations in each case – and in addition, and importantly, they are shown to be local in their influence. During map construction the user's map commands are stored in a log file which can be referenced for map re-creation or roll-back. A temporal tree structure may be maintained showing the history of the changes to each map entity.

FURTHER READING

Maps change. Even apart from kinetic maps, as described previously under 'Marine GIS', roads and buildings are added and removed from the landscape: events happen. Langren (1992) made one of the early reviews. Peuquet (1999, 2001) explored time in GIS and spatial analysis. Gold (1996) focussed on the event-driven approach and Mioc et al. (1999, 2012) examined the algebra of a Voronoi-event model. This can be particularly useful in the job of maintaining historical cadastral maps.

REFERENCES

Ahuja, N. & Tuceryan, M. (1989) Extraction of early perceptual structure in dot patterns: Integrating region, boundary and component Gestalt. *Computer Vision, Graphics and Image Processing*, 48, 304–356.

Amenta, N., Bern, M. & Eppstein, D. (1998) The crust and the beta-skeleton: Combinatorial curve reconstruction. *Graphical Models and Image Processing*, 60, 125–135.

Anton, F., Snoeyink, J. & Gold, C.M. (1998) An iterative algorithm for the determination of Voronoi vertices in polygonal and non-polygonal domains on the plane and the sphere. In: *14th European Workshop on Computational Geometry (CG'98). Barcelona, Spain.*

Blum, H. (1967) A transformation for extracting new descriptors of shape. In: Whaten-Dunn, W. (ed.) *Models for the Perception of Speech and Visual Form.* Cambridge, MA, M.I.T. Press. pp. 153–171.

Blum, H. (1973) Biological shape and visual science (Part I). *Journal of Theoretical Biology*, 38, 205–287.

Blum, H. & Nagel, R.N. (1978) Shape description using weighted symmetric axis features. *Pattern Recognition*, 10, 167–180.

Boots, B.N. (1974) Delaunay triangles: An alternative approach to point pattern analysis. *Proceedings, Association of American Geographers*, 6, 26–29.

Burnikel, C., Melhorn, K. & Schirra, S. (1994) How to compute the Voronoi diagram of line segments: Theoretical and experimental results. In: *Lecture Notes in Computer Science, 855 (Proceedings of the Second Annual European Symposium on Algorithms ESA '94).* Berlin, Springer-Verlag. pp. 227–239.

Chew, L.P. (1987) Constrained Delaunay triangulations. In: *Proceedings of the Third Annual Symposium on Computational Complexity.* pp. 215–222.

Chew, L.P. (1989) Constrained Delaunay triangulations. *Algorithmica*, 4, 97–108.

Christaller, W. (1933) Die Zentralen Orte in Suddeutschland Jena: Fisher. (C. Baskin, Trans. *The Central Places of Southern Germany.* Englewood Cliffs, Prentice-Hall, 1966).

Christensen, A.H.J. (1987) Fitting a triangulation to contour lines. In: *Proceedings of AutoCarto.* Vol. 8. pp. 57–67.

Dakowicz, M. & Gold, C.M. (2003) Extracting meaningful slopes from terrain contours. *International Journal of Computational Geometry and Applications*, 13, 339–357.

Dakowicz, M. & Gold, C.M. (2006) Structuring kinetic maps. In: Reidl, A., Kainz, W. & Elmes, G. (eds.) *Progress in Spatial Data Handling-12th International Symposium on Spatial Data Handling.* Berlin, Springer. pp. 477–493.

Dakowicz, M. & Gold, C.M. (2007) Finite difference method runoff modelling using Voronoi cells. In: *Proceedings: 5th ISPRS Workshop on Dynamic and Multi-dimensional GIS, Urumchi, China.* pp. 55–60.

de Berg, M., van Oostrum, R. & Overmars, M. (1996) Simple traversal of a subdivision without extra storage. In: *Proceedings, 12th Annual Symposium on Computational Geometry.* pp. C5–C6.

de Berg, M., van Kreveld, M., van Oostrum, R. & Overmars, M. (1997) Simple traversal of a subdivision without extra storage. *International Journal of Geographical Information Science*, 11, 359–374.

De Floriani, L. (1989) Two algorithms for radial ordering of Delaunay triangles. In: *Abstracts of the First Canadian Conference on Computational Geometry.* Vol. 40. Montreal Quebec.

De Floriani, L. & Puppo, E. (1992) An on-line algorithm for constrained Delaunay triangulation. *Computer Vision, Graphics and Image Processing*, 54, 290–300.

De Floriani, L., Falcidieno, B., Nagy, G. & Renovi, C. (1991) On sorting triangles in a Delaunay triangulation. *Algorithmica*, 6, 522–532.

De Floriani, P. & Magillo, P. (2003) Algorithms for visibility computation on terrains: A survey. *Environment and Planning B – Planning and Design*, 30, 709–728.

Devillers, O., Pion, S. & Teillaud, M. (2002) Walking in a triangulation. *International Journal of Foundations of Computer Science*, 13, 181–199.

Dutton, G. (1996) Encoding and handling geospatial data with hierarchical triangular meshes. In: Kraak, M.J. & Molenaar, M. (eds.) *Advances in GIS Research II: Proceedings, Spatial Data Handling.* Vol. 7. London, Taylor & Francis. pp. 505–518.

Edla, D.R. (2003) Developing algorithms for clustering biological data – Chapter 4: Voronoi diagram based clustering algorithms. PhD Thesis. Indian School of Mines, 144 pp.

Estivill-Castro, V. & Lee, I. (1999) AMOEBA: Hierarchical clustering based on spatial proximity using Delaunay diagram. In: *Proc. 9th Int. Symp. Spatial Data Handling (SDH'99).* pp. 7a.26–7a.41.

Estivill-Castro, V. & Lee, I. (2000) AUTOCLUST: Automatic clustering via boundary extraction for mining massive point-data sets. In: Abrahart, J. & Carlisle, B.H. (eds.) *Proc. 5th Int. Conf. on Geocomputation.* pp. 23–25.

Fuchs, H., Kedem, Z.M. & Uselton, S.P. (1977) Optimal surface reconstruction from planar contours. *Communications of the ACM*, 20, 693–702.

Gold, C.M. (1990) Spatial data structures – The extension from one to two dimensions. In: Pau, L.F. (ed.) *Mapping and Spatial Modelling for Navigation*. NATO ASI Series F No. 65. Berlin, Springer-Verlag. pp. 11–39.

Gold, C.M. (1992) The meaning of 'neighbour'. In: *Lecture Notes in Computing Science 639: Theories and Methods of Spatio-Temporal Reasoning in Geographic Space*. Berlin, Springer-Verlag. pp. 220–235.

Gold, C.M. (1996) An event-driven approach to spatio-temporal mapping. *Geomatica*, 50, 415–424.

Gold, C.M. (1997) Simple topology generation from scanned maps. In: *Proceedings of Auto-Carto*. Vol. 13 (5). pp. 337–346.

Gold, C.M. (1999) Crust and anti-crust: A one-step boundary and skeleton extraction algorithm. In: *Proceedings of the Fifteenth Annual Symposium on Computational Geometry*. pp. 189–196.

Gold, C.M. (2010) The dual is the context: Spatial structures for GIS. In: *Proceedings, 2010 International Symposium on Voronoi Diagrams in Science and Engineering*. IEEE. pp. 3–10.

Gold, C.M. & Angel, P. (2006) Voronoi hierarchies. In: Raubal, M., Miller, H., Frank, A. & Goodchild, M.F. (eds.) *Geographic, Information Science. Lecture Notes in Computer Science*. Springer Berlin/Heidelberg. pp. 99–111.

Gold, C.M. & Cormack, S. (1987) Spatially ordered networks and topographic reconstructions. *International Journal of Geographical Information Systems*, 1, 137–148.

Gold, C.M. & Dakowicz, M. (2005) The crust and skeleton – Applications in GIS. In: *Proceedings, 2nd International Symposium on Voronoi Diagrams in Science and Engineering*. pp. 33–42.

Gold, C.M. & Dakowicz, M. (2006) Kinetic Voronoi/Delaunay drawing tools. In: *Proceedings 3rd International Symposium on Voronoi Diagrams in Science and Engineering (ISVD 2006)*. pp. 76–84.

Gold, C.M. & Maydell, U.M. (1978) Triangulation and spatial ordering in computer cartography. In: *Proceedings, Canadian Cartographic Association Third Annual Meeting*. pp. 69–81.

Gold, C.M. & Snoeyink, J. (2001) A one-step crust and skeleton extraction algorithm. *Algorithmica*, 30, 144–163.

Gold, C.M. & Thibault, D. (2001) Map generalization by skeleton retraction. In: *Proceedings, 20th Int. Cartographic Conference ICC 2001*. International Cartographic Association. pp. 2072–2081.

Gold, C.M., Charters, T.D. & Ramsden, J. (1977) Automated contour mapping using triangular element data structures and an interpolant over each triangular domain. In: George, J. (ed.) *Proceedings Siggraph '77. Computer Graphics*. Vol. 11. pp. 170–175.

Gold, C.M., Nantel, J. & Yang, W. (1996) Outside-in: An alternative approach to forest map digitizing. *International Journal of Geographical Information Systems*, 10, 291–310.

Gold, C.M., Remmele, P.R. & Roos, T. (1995) Voronoi diagrams of line segments made easy. In: *Proceedings of the 7th Canadian Conference on Computational Geometry*. pp. 223–228.

Gower, J.C. & Ross, G.J.S. (1969) Minimum spanning trees and single linkage cluster analysis. *Journal of the Royal Statistical Society Series C*, 18, 54–64.

Held, M. (2001) VRONI: An engineering approach to the reliable and efficient computation of Voronoi Diagrams of points and line segments. *Computational Geometry, Theory and Application*, 18, 95–123.

Imai, T. & Sugihara, K. (1994) A failure-free algorithm for constructing Voronoi diagrams of line segments. *Transactions of the Information Processing Society of Japan*, 35, 1966–1977.

Karavelas, M.I. (2004) A robust and efficient implementation for the segment Voronoi Diagram. In: *International Symposium on Voronoi Diagrams in Science and Engineering 2004*. pp. 51–62.

Kimia, B. (2004) On the role of medial geometry in human vision. *Journal of Physiology-Paris*, 97 (2–3), 155–190.

Langran, G. (1992) *Time in Geographic Information Systems*. London, Taylor & Francis.

Lardin, P. (1999) Application de la structure des données Voronoi à la simulation de l'écoulement des eaux souterraines par différences finies intégrées. Mémoire de maîtrise en sciences géomatiques, Faculté de foresterie et de géomatique, Université Laval, Québec. 159 pp.

Lee, I. & Estivill-Castro, V. (2002) Polygonization of point clusters through cluster boundary extraction for geographical data mining. In: *Advances in Spatial Data Handling, 10th International Symposium on Spatial Data Handling*. Berlin, Springer. pp. 27–40.

Matuk, K., Gold, C.M. & Li, Z. (2006) Skeleton based contour line generalization. In: Kainz, W., Reidl, A. & Elmes, G. (eds.) *Progress in Spatial Data Handling – 12th International Symposium on Spatial Data Handling*. Berlin, Springer. pp. 643–658.

Mioc, D., Anton, F., Gold, C.M. & Moulin, B. (1999) "Time travel" visualization in a dynamic Voronoi data structure. *Cartography and Geographic Information Science*, 26, 99–108.

Mioc, D., Anton, F., Gold, C.M. & Moulin, B. (2012) Map updates in a dynamic Voronoi data structure. In: Monwar Alam, B. (ed.) *Application of Geographic Information Systems*. InTech. pp. 37–64.

Narasimhan, T.N. & Witherspoon, P.A. (1976) An integrated finite difference method for analyzing fluid flow in porous media. *Water Resources Research*, 12 (1), 57–64.

Ng, R.T. & Han, J. (1994) Efficient and effective clustering method for spatial data mining. In: Bocca, J.B., Jarke, M. & Zaniolo, C. (eds.) *Proc. of the 20th Int. Conf. on Very Large Data Bases*. Morgan Kaufmann. pp. 144–155.

Ogniewicz, R.L. (1992) *Discrete Voronoi Skeletons*. PhD Thesis. Zurich, Switzerland, Swiss Federal Institute of Technology.

Ogniewicz, R.L. & Ilg, M. (1992) Voronoi skeletons: Theory and applications. In: *Proceedings of the 1992 IEEE Computer Society Conference on Computer Vision and Pattern Recognition*. pp. 63–69.

Peuquet, D.J. (1999) Time in GIS and geographical databases. In: Longley, P.A., Goodchild, M.F., Maguire, D.J. & Rhind, D.W. (eds.) *Geographical Information System, Principles and Technical Issues*. 2nd edition. Vol. 1. New York, NY, John Wiley & Sons. pp. 91–103.

Peuquet, D.J. (2001) Making space for time: Issues in space-time data representation. *GeoInformatica*, 5, 11–32.

Rohlb, F.J. (1973) Hierarchical clustering using the minimum spanning tree. *The Computer Journal*, 16, 93–95.

Sang, N., Gold, C.M. & Miller, D. (2007) Shadow of the Stanford Bunny: Analysing visual models of landscape data using Delaunay TIN. In: *4th International Symposium on Voronoi Diagrams*. pp. 294–299.

Snoeyink, J. (1997) Point location. In: Goodman, J.E. & O'Rourke, J. (eds.) *Handbook of Discrete and Computational Geometry*. Boca Raton, FL, CRC Press. pp. 559–574.

Streit, U. & Wiesmann, K. (1996) Problems of integrating GIS hydrological models. In: Fischer, M., Scholten, J.H. & Unwin, D. (eds.) *Spatial Analytical Perspectives on GIS*. London, Taylor & Francis. pp. 161–175.

Sui, D.Z. & Maggio, R.C. (1999) Integrating GIS with hydrological modeling: Practices, problems, and prospects. *Computers, Environment and Urban Systems*, 23, 33–51.

Thibault, D. & Gold, C.M. (2000) Terrain reconstruction from contours by skeleton construction. *GeoInformatica*, 4, 349–373.

Tuceryan, M. & Jain, A.K. (1990) Texture segmentation using Voronoi polygons. *IEEE Transactions on Pattern Analysis and Machine Intelligence*, 12, 211–216.

van Oosterom, P. (1999) Spatial access methods. In: Longley, P.A., Goodchild, M.F., Maguire, D.J. & Rhind, D.W. (eds.) *Geographical Information Systems*. Vol. 1. New York, NY, John Wiley & Sons. pp. 385–400.

Walters, R.F. (1969) Contouring by machine: A users' guide. *American Association of Petroleum Geologists Bulletin*, 53, 2324–2340.

Wang, C.A. & Schubert, L.K. (1987) An optimal algorithm for constructing the Delaunay triangulation of a set of line segments. In: *Proceedings of the Third ACM Symposium on Computational Geometry*. pp. 223–232.

Yang, W. & Gold, C.M. (1995) Dynamic spatial object condensation based on the Voronoi diagram. In: *Proceedings, Fourth International Symposium, LIESMARS'95 – Towards Three-Dimensional, Temporal and Dynamic Spatial Data Modelling and Analysis. Wuhan, China*.

Yang, W. & Gold, C.M. (1996) Managing spatial objects with the VMO-Tree. In: Kraak, M.J. & Molenaar, M. (eds.) *Seventh International Symposium on Spatial Data Handling, Delft, The Netherlands*. Vol. 11B. pp. 15–30.

Zahn, C.T. (1971) Graph-theoretical methods for detecting and describing gestalt clusters. *IEEE Transactions on Computers, C*, 20, 68–86.

Chapter 5

2D GIS

5.1 SPATIAL DECISION SUPPORT SYSTEMS

Originally what is now 'GIS' was 'Automated Cartography', and its main purpose was the drawing of maps. Early GIS was similar, but with basic analytic functions, such as labelling and merging polygons, and drawing buffer zones. The earliest accepted GIS was a raster system for estimating the Canadian wheat harvest – see Tomlinson (1967).

The emphasis was largely on digitizing and constructing polygon maps, and overlaying different layers, for example to find how much top grade agricultural land would be eliminated by new urban zoning. The underlying layers, however, were static. This allowed some limited 'What if . . .' queries.

Other data types required different data structures, e.g. for grid elevation models, for TIN models, for building outlines and for road networks. This made rapid comparison of combinations of layers difficult to achieve, because the data structures were primarily static, needing to be rebuilt completely after any changes – and the incompatibility of different layers made their combination particularly difficult.

However, with the idea of a 'Spatial Decision Support System' (SDSS) came a perception that the human-computer interface was highly significant. Here the purpose was the assistance to the operator to make near-optimal decisions for queries that were unlikely to be able to be framed formally: an example might be the selection of forest harvesting roads, that involved an evaluation of the trafficability of the terrain, easy access to high quality forest stands, and concern for the effects of cutting on both wildlife migration and destruction of important scenic areas.

This type of analysis depends on close-knit interaction between human and computer. Computers are very good at fast, well defined calculations, while humans are good at rapid approximate solutions. Humans also have a very powerful short-term memory, capable of retaining amorphous half-formed views and images – but it fades within a couple of minutes if not refreshed. Thus there needs to be a fast feedback loop, with the computer displaying some partial solution (perhaps to the forest road problem just mentioned) – a 'scenario' – and the human operator (a forest manager who understands his land) criticising it and suggesting modifications which can be entered rapidly into the computer – which has to respond with a hopefully improved scenario within approximately two minutes, before the operator has forgotten the image that motivated him. The interactive component of an SDSS is intended to allow the user to examine the consequences of a particular scenario and then modify it while his 'idea' is still fresh in memory. See Gold (1993).

This requires redesign of spatial analysis systems from the bottom up to allow rapid data and data structure modification, and fast simulation techniques. These features are not available in most spatial analysis systems. Our experience in forestry suggests that they give significant financial advantages where the user's experience is a necessary part of the scenario specification process.

An initial scenario is prepared, maybe manually, that approximates the type of solution the manager expects: he probably envisages a road from the mill extending towards the highest value stands, for example. The proposed features need to be added locally, using easily understood operations to integrate the changes with the existing map information. This is submitted to the computer system, which then evaluates the modified scenario according to the specified needs: the volume of harvestable wood within 200 m of the road, for example, the slope and soil type (swamps are not recommended), un-harvested gaps for wildlife movement, and visibility from public roads.

This needs to be evaluated and returned to the manager, either as simple totals or graphical modifications of the map, before he has lost track of his original idea: maybe on the side-slope below the crest, for example, hiding it from the public road and linking two valuable stands with a gap between. He examines the result and either nods or shakes his head, and probably makes further modifications – either major or minor – before re-submitting it. In the end – perhaps after several days – he will accept the result as the best he can get at present, and move on.

The same approach can be used in a 'war-room' setting, with several experts gathered round a large screen: this may be a high-level meeting, or a public hearing where public objections and suggestions can be tested in real time, and shown to be feasible, impractical or have serious side-effects. It should be noted that the results from this approach can in no way be expected to be a global optimum – which would require evaluation of all possible scenarios – but a reasonable solution equivalent to a manual evaluation that would take considerably longer. A different manager, starting at a different point, might well have arrived at a different solution.

The technical requirements for such a system differ significantly from the traditional approach. The geometric layer and its context layer – the DT/VD – must be editable interactively and in real-time. Ideally they should be seamless, and should not need to be sliced into small pieces along geographical coordinates. The different layers – objects, networks, polygons, terrain – need to be integrated in real-time, so ideally they should use a consistent data structure: we need a 'Unified Spatial Model'. While the dual may always be derived from the primal data structure, and vice versa, either structure may be needed, or both, for a particular layer.

The most common data types in a 2D GIS are discrete objects (points, building outlines), networks (rivers, roads), polygons (choropleth maps, soil types) and surfaces or fields (TIN models, smooth interpolated models). We have shown previously how most of these may be constructed using points and line segments linked by the DT/VD structure. They may thus be stored, modified and analysed using a common set of operations, rather than different ones for each application. Upon completion, the differing geographic layers may be combined in 'polygon overlay' operations using the same processes: while not all combinations may be geographically meaningful, the technical operations are consistent, and can produce new and interesting overlays.

5.2 THE UNIFIED SPATIAL MODEL

5.2.1 *Discrete objects: points*

The simple VD of a point set is basic to many operations, especially various types of surface modelling. The Voronoi neighbours are consistently defined by the InCircle test during the diagram construction, and the dual Delaunay triangulation edges connect the generators of pairs of adjacent Voronoi cells. The insertion or deletion of the central point only affects the neighbours, showing that the diagram may be updated locally, and 'area-stealing' interpolation uses only these neighbours. Boundaries between points with differing labels may be used as a preliminary mapping of various data types.

Strings of points may have a particular significance, perhaps as digitized polygon, contour or river boundaries. Given sufficiently dense point spacing, the crust/skeleton test, using InCircle, may be used to delineate the boundaries using the DT, and the skeleton separators, using the VD.

5.2.2 *Impermeable linear features*

Man-made artefacts often have long straight-line segments for their boundaries. This is also true for digitized features inserted into a terrain model, such as ridges and valleys. A common approach to handling this is the Constrained DT (Figure 234) where some of the triangle edges (green) represent boundaries – the connections between corners – and some (black) merely represent Delaunay adjacency. As discussed before, this mixes the entity layer and the adjacency layer, with points in the first and edges in the second. This reduces the number of types of spatial query that may be performed, and limits ready database access.

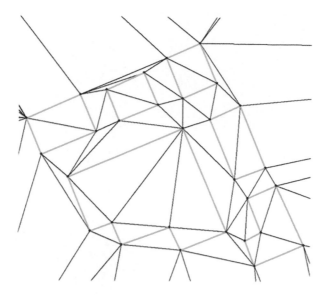

Figure 234.   CDT model of some buildings.

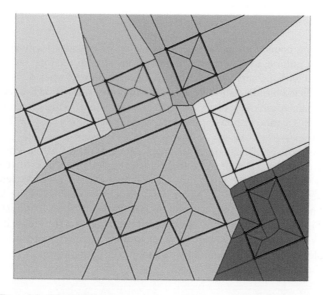

Figure 235.   LSVD model of the buildings.

When both points and line segments may be stored in the entity layer, with the VD/DT stored separately in the adjacency layer, then context queries may be made of all the map objects, for such properties as adjacency, separators, interior and exterior spatial extensions, and navigation (Figure 235).

5.2.3   *Skeleton: grouped contexts*

The Voronoi cell is the extension of a map object, or a collection of contiguous objects. As line segments have sidedness, each half-line has its own extension – for example the interior or the

151

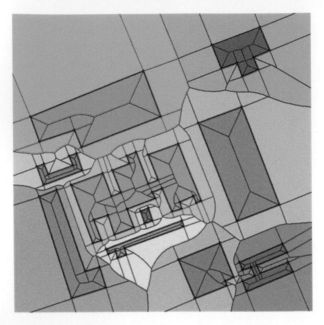

Figure 236.   LSVD and grouped contexts.

exterior of a building boundary's extension. A collection of boundary segments and points may form a 2D building outline, and their interior extensions sum to the total area of the building – and may be labelled with the appropriate attributes of the building: when a 'What is here?' query is asked, the result is the building attribute. The exterior extensions form the region closest to the building, and return the result of the query 'What is closest to here?' Thus even for a map of discrete objects there is always a valid response – meaning that the set of 'objects' has been converted to a 'field' (Figure 236).

The grouped contexts, or extensions, of the map objects form the extensions of the completed buildings, and the resulting skeletons indicate the adjacency relationships between the buildings – which are next to which, as in the village model sketched earlier. These skeletons also map the navigability between the buildings: the Voronoi nodes are the centres of the largest empty circumcircles in the neighbourhood, and the resulting boundaries show the equidistant lines between pairs of objects. Finding whether a large (circular) object, such as a fire truck, can navigate to the scene of the fire is therefore directly obtained by examining the circumcircles along the path, and the set of available edges may be explored with any of the graph traversal techniques described earlier.

5.2.4   *Buffer: clipped contexts*

Instead of asking about the widest available path, the set of paths with a pre-specified width may be obtained instead. As the Voronoi cells define the regions closest to each object, calculation of buffer zones may be phrased as: 'Within each object's extension, clip off the region within the desired distance.' This may be achieved with linear segments or circular arcs (Figure 237).

5.2.5   *Overlays*

Voronoi layers may be conflated to produce combined overlays, whether for multiple polygon sets, collections of discrete objects, contour maps or networks. Figure 238 shows a building layer, with the Voronoi cells for all map objects (and object parts).

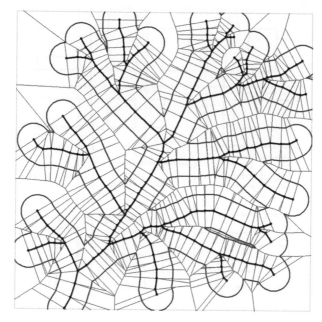

Figure 237.   Clipped contexts: buffer zones.

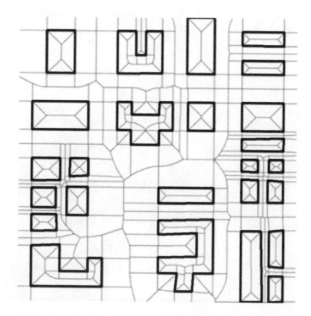

Figure 238.   Urban LSVD.

Figure 239 combines the building layer with a road layer. Subsequent analysis may determine if sufficient road width is available for proposed future construction.

Polygon maps may be handled similarly (Figure 240). Line segments along the boundaries will have extensions on each side, each one associated with the appropriate polygon – 'voids' in our Cosmic Model. Boundaries are by nature double, and here a comparison with the earlier scanned-map examples is informative: both are supporting the same spatial model, with two crusts or

153

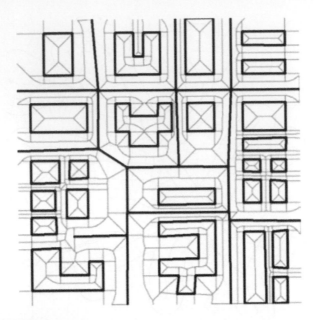

Figure 239.   Urban LSVD plus roads.

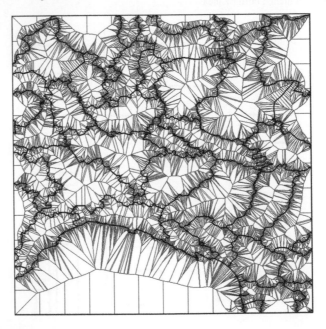

Figure 240.   Polygon map LSVD.

half-lines separated by a skeleton – in the case of the line segments the skeleton, or extension boundary, is down the middle of the line, separating the two halves – think razor blades!

5.2.6 *Networks*

Road networks may be handled in the same way as polygons, as the polygon boundaries above could well be imagined as roads, with fields on either side. Several advantages result from this approach.

Figure 241.   Road network LSVD.

Firstly topological network errors where line segments fail to meet, or intersect, do not cause topological errors in the underlying map: graph queries of the type discussed above may be used to detect small gaps or very short intersecting segments. There is no need to rebuild the whole network to fix errors, as is the case with traditional one-dimensional arcs: a navigable map is always available and local searches within it are always valid. Thus minor connectivity errors of the map objects – the roads – may be fixed locally, and often automatically.

Secondly, the resulting roads are by nature bi-directional, composed of half-lines, allowing two-way traffic: the graph is already prepared for navigability queries, traffic flow and travel time analysis, perhaps with the addition of simple segment attributes such as distance or road condition.

Thirdly, each half-line has its spatial extension associated with it, meaning that origins and destinations need not fall exactly on the one-dimensional road element. In Figure 241 the location 'x' falls within the extension of a particular half-line segment, and queries may start, or end, at that segment: a fire truck may answer a response to a field fire without additional analysis!

Rivers follow exactly the same process. Hydrography segments are linked in the same way as roads, and the skeletons (watersheds) and Voronoi cells (minor catchment areas) follow automatically, as described previously. Given the graph structure underpinning the whole map, basic runoff and flow may be estimated directly, as described before.

5.2.7   Combining them all

The resulting Unified Spatial Model may thus be outlined (Figure 242). Different types of discrete objects, networks, polygons and terrain models may be combined using the same structure, and analysed with similar tools and algorithms. In addition, new combinations may arise: there is no difficulty having additional points, line segments or objects within a polygon for example, or adding terrain features to a city map.

5.2.8   Mobile points

Finally, the kinetic model may be added, to permit collision detection and navigation within any of the resulting maps. In Figure 243 points a, b, c and d are free-moving vehicles, or robots, within

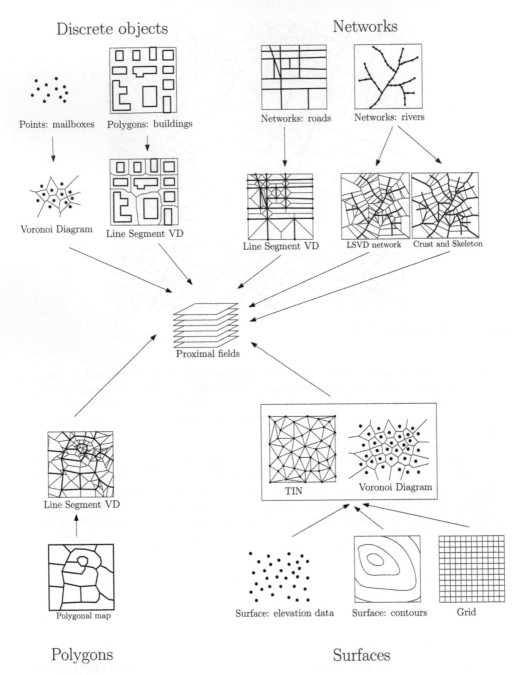

Figure 242. Unified Spatial Model.

the city. Their spatial extents indicate the adjacent buildings to be avoided, and in the cases of c and d the potential for inter-vehicle collision may be evaluated – just as we saw in the Marine GIS discussed previously.

The ability to combine different data types, and analyze and modify them in parallel provides the underlying framework for a true real-time SDSS, as exemplified by a forest harvesting DSS

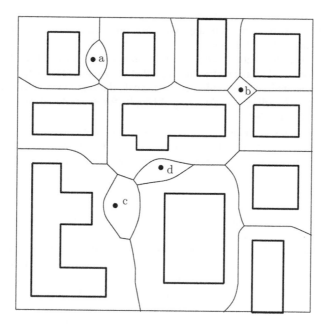

Figure 243. Moving objects in LSVD.

(as illustrated in Figure 242), a marine GIS (as described under the section about kinetic point modelling), or a historical cadastral mapping system (using the kinetic line segment updating model) – and many other applications.

FURTHER READING

Tomlinson (1967) designed the first modern GIS, based on a raster system that was intended as a decision support system. Gold (1993) described the structure of an interactive (flight simulator) approach to decision making. Dakowicz and Gold (2012) described the 'Unified Spatial Model' where both the topological structuring and object specification could be modified locally and interactively while preserving all spatial relationships.

The key issue here was the preservation of all topological structuring with an interactive topology editor – based on the dual Voronoi/Delaunay diagrams. Corbett (1975) and Chrisman (1975) discussed the underlying static topological structure of polygon boundaries. Boudriault (1987) described the linked boundary segments used in the US Census maps.

There were extensive discussions on the vector/raster distinction, resulting in the object/field distinction for geographic entities. Examples of this are the work of Couclelis (1992) and Cova and Goodchild (2002), with a good summary by Mark (1999) and reviews of the Voronoi contribution by Gold (2006, 2010a, 2010b, 2016 in press).

5.3   OTHER VORONOI TECHNIQUES

We have mostly discussed the VD/DT on the Euclidean plane – while, however, admitting that the sphere gives a conceptually pleasing continuity to the topology of the boundaries. The circumcentre calculation with the vector normal is also simple and elegant, but for geographic data the number of digits of precision required is high – for example to handle a ground spacing of less than a metre on a framework with the origin at the centre of the earth – with an equatorial radius of approximately

6,378 km and a flattening at the poles of approximately one part in 273. See the work of Lukatela (1989) for calculations on the ellipsoid.

Nevertheless, the Voronoi spatial model is applicable on more than just the Euclidean plane or sphere. It can be constructed on any metric surface – that is, any surface where the distance between two points can be calculated, such as a cylinder – as the calculations depend on the concept of spatial dominance, as described earlier. It is thus very difficult, for example, to model the Voronoi-like pattern on the skin of a giraffe.

However, it is feasible to calculate the Voronoi diagram on a graph, such as for a road network, perhaps to calculate the zones closest to a hospital or fire station, where the shortest travel time may be calculated between any two network points, by travelling exclusively along the roads. See Okabe et al. (2000). For the hospital-districting problem this is probably more realistic than simply using the 2D Euclidean space model.

So far we have been discussing Voronoi methods based on an object model – where each object is represented by one or more coordinate pairs. We can also examine spatial relationships within a field model of space, such as a raster.

The first thing to note about a field model is that it is not an object model – there are no explicitly defined objects, only attributes associated with particular grid squares or other polygons. This means that recovering the spatial extent of an object involves scanning a large number of pixels, and then reconstructing its outline – a raster-to-vector conversion process that has a variety of difficulties that are well discussed in the literature.

A second difficulty is that, due to the orientation of the field subdivision – which way the grid is oriented – any need to work with a grid at a different orientation will generate a different Voronoi diagram that may not have the same connectivity as the original. For many applications this may not be a problem, but sometimes it can give inconsistent results.

Finally, it is important to realise that in a raster distance measurements are made in a different metric: it uses L^1 (city block) metric, which means that a circle (a boundary equidistant from a central point) looks like a diamond. Thus the VD in L^1 gives a very different set of neighbours from a VD in L^2 (Euclidean distance). Various attempts have been made to simulate L^2 in an L^1 space, with varying degrees of success – simulating an L^2 circle is not particularly easy, and the L^2 VD requires the InCircle test. However, circle growing algorithms can approximate some of these issues satisfactorily for many applications. See Li et al. (1999).

Nevertheless, various applications can be satisfactorily resolved with a raster approach – especially where the 'objects' are complex, perhaps too complex to be handled effectively with the Euclidean object model. We will mention several examples and compare between the object and field models, remembering that in many ways the Voronoi approach is not an algorithm, but a hybrid model of space with both object and field properties. This becomes particularly relevant when complex objects are captured as raster images and we wish to calculate the Voronoi regions of a set of houses, for example. This can only be done on a pixel-by-pixel basis, usually with two sweeps through the raster for each pixel, as described in Li et al. (1999) and others, giving a distance labelling of each pixel – the result is an increasing ring of distances from each object, and the line of the maximum distances – the watershed – gives the Voronoi boundary. Chen et al. (2004a and b) used these techniques to construct the contour tree of a terrain model based on the adjacency of the Voronoi regions of each contour string. Fabbri et al. (2008) gives a recent survey. Chen et al. (2001) used the raster VD to extend the '9-Intersection Model' of Egenhofer and Franzosa (1991) of the possible combinations of the interior, boundary and exterior of two possibly intersecting objects – by examining the relationships of the objects' Voronoi regions as well. This is used, for example, when recovering individual objects from a database and testing their topological relationships.

Because 2D geographic space is often embedded in the sphere or ellipsoid, various workers have attempted to implement the VD there. Chen et al. (2003) generated the VD on the QTM partition of the sphere using triangle dilation/erosion operators, for 'raster' type objects, and Hu, Liu and Hu (2014) more recently constructed the Voronoi diagram on the ellipsoid using the distance transform approach, making some geographic Voronoi problems more tractable.

In the planar Euclidean object model there is the work of Martin Held – a particularly robust implementation that is often used for elaborate router cutting of CAD parts: this is particularly interesting as its application is based on Blum's wave-front analogy for generating the VD – and is also related to Christensen's cartographic application. Held (2001) was able to construct the Voronoi diagrams of complex objects directly in Euclidean space. His work was mainly aimed at offset machining, where a router is used to cut any arbitrary pocket – which is equivalent to expanding interior wave-fronts of a polygon. His work is the result of many years of development and appears to be extremely robust – and robustness is particularly difficult to achieve in problems of this complexity. Thus, depending on the application, various Voronoi methods are available and useful for complex mapping applications.

FURTHER READING

Researchers into other Voronoi approaches includes Lukatela (1989), who described a spherical point Voronoi indexing system for engineering data, as did Chen et al. (2003). More recently Hu et al. (2014) described the point VD for the ellipsoidal earth.

Chen et al. (2001, 2004a, 2004b) refined the Egenhofer and Franzosa (1991) point-set topology. Li et al. (1999) developed raster-based Voronoi techniques, and Fabbri et al. (2008) reviewed many related distance-transform methods. The basic reference of Okabe et al. (2000) includes many other GIS and Voronoi methods.

REFERENCES

Boudriault, G. (1987) Topology in the TIGER file. In: *Proceedings 8th International Symposium on Computer Assisted Cartography*. pp. 258–269.

Chen J., Li, C., Li, Z. & Gold, C.M. (2001) Voronoi-based 9-intersection model for spatial relations. *International Journal of Geographical Information Science*, 15, 201–220.

Chen, J., Qiao, C. & Zhao, R. (2004a) A Voronoi interior adjacency-based approach for generating a contour tree. *Computers & Geosciences*, 30, 355–367.

Chen, J., Zhao, R. & Li, Z. (2004b) Voronoi-based k-order neighbour relations for spatial analysis. *ISPRS Journal of Photogrammetry and Remote Sensing*, 59, 60–72.

Chen, J., Zhao, X. & Li, Z. (2003) An algorithm for the generation of Voronoi diagrams on the sphere based on QTM. *Photogrammetric Engineering and Remote Sensing*, 69, 79–89.

Chrisman, N.R. (1975) Topological information systems for geographic representation. In: *Proceedings of the Second International Symposium on Computer-Assisted Cartography (Auto-Carto 2)*. ASPRS/ACSM. pp. 346–351.

Corbet, J.P. (1975) Topological principles in cartography. In: *Proceedings, Auto-Carto II*. pp. 61–65.

Couclelis, H. (1992) People manipulate objects (but cultivate fields): Beyond the raster-vector debate in GIS. In: *Theories and Methods of Spatio-Temporal Reasoning in Geographic Space*. pp. 65–77.

Cova, T.J. & Goodchild, M.F. (2002) Extending geographical representation to include fields of spatial objects. *International Journal of Geographic Information Science*, 16, 509–532.

Dakowicz, M. & Gold, C.M. (2012) A unified spatial model for GIS. In: Shi, W., Goodchild, M.F., Lees, B. & Leung, Y. (eds.) *Advances in Geo-Spatial Information Science*. Boca Raton, FL, CRC Press. pp. 13–24.

Egenhofer, M.J. & Franzosa, R.D. (1991) Point-set topological spatial relations. *International Journal of Geographical Information Systems*, 5, 161–176.

Fabbri, R., Costa, L., Da F., Torelli, J.C. & Bruno, O.M. (2008) 2D Euclidean distance transform algorithms: A comparative survey. *ACM Computing Surveys*, 40, 1.

Gold, C.M. (1993) Forestry spatial decision support system classification, and the 'flight simulator' approach. In: *Proceedings GIS'93: Eyes on the Future*. pp. 797–802.

Gold, C.M. (2006) What is GIS and what is not? *Transactions in GIS*, 10, 505–519.

Gold, C.M. (2010a) The dual is the context: Spatial structures for GIS. In: *Proceedings, 2010 International Symposium on Voronoi Diagrams in Science and Engineering (ISVD 2010)*. pp. 3–10.

Gold, C.M. (2010b) Voronoi diagrams. In: Warf, B. (ed.) *Encyclopaedia of Geography*. California, SAGE Publications. pp. 3033–3035.

Gold, C.M. (2016, in press) The Voronoi Data structure. In: *The International Encyclopedia of Geography: People, the Earth, Environment, and Technology*. Chichester, John Wiley & Sons Ltd.

Hu, H., Liu, X. & Hu, P. (2014) Voronoi diagram generation on the ellipsoidal earth. *Computers & Geosciences*, 73, 81–87.

Li, C., Chen, J. & Li, Z. (1999) A raster-based method for the generation of Voronoi diagrams for spatial objects using dynamic distance transformation. *International Journal of Geographic Information Science*, 13, 209–225.

Lukatela, H. (1989) Hipparchus data structure: Points, lines and regions in a spherical Voronoi grid. In: *Proc. Ninth International Symposium on Computer-Assisted Cartography (Auto-Carto 9)*. pp. 164–170.

Mark, D.M. (1999) Spatial representation: A cognitive view. In: Longley, P.A., Goodchild, M.F., Maguire, D.J. & Rhind, D.W. (eds.) *Geographical Information Systems*. 2nd edition. New York, NY, John Wiley and Sons. pp. 81–89.

Okabe, A., Boots, B., Sugihara, K. & Chiu, S.N. (2000) *Spatial Tessellations – Concepts and Applications of Voronoi Diagrams*. 2nd edition. Chichester, John Wiley and Sons. 671 pp.

Tomlinson, R.F. (1967) *An Introduction to the Geographic Information System of the Canada Land Inventory*. Ottawa, ON, Department of Forestry and Rural Development.

Chapter 6

3D GIS

6.1   3D BOUNDARY FACES

6.1.1   *Embedded Spaces*

We now need to look at other spatial dimensions than the planar map. In GIS we often confuse the dimension of the space we are thinking of, with the dimension of the objects or structures. We have discussed data structures previously. We will now look at how they embed in different spatial dimensions.

We can embed objects in a particular spatial environment: e.g. points on a line, or a grid on a sphere. These objects (structures) must be of the same or lower dimension than the embedding space. However, the same structure type may have different meanings in different spaces, as seen below:

Space-dimensions						
		0	1	2	3	Data type:
Embeded-	0	F	B	BB	BBB	nodes
Dimension	1		F	B	BB	graphs
	2			F	B	surfaces
	3				F	volumes

The symbols mean:

F – The space is 'filled' with structures of the same dimension. This could be a field covering the whole map, or smaller regions.
B – Boundaries of the filled region
BB – Boundaries of boundaries (e.g. nodes on polygon boundaries)
BBB – One level more (e.g. nodes on edges of faces of 3D solids)

Thus in two spatial dimensions a 2D map entity occupies all the available degrees of freedom (F), its 1D boundary graph surrounds it, and each edge of it is itself bounded by 0D nodes. The same would be true for a 2D triangulation in 2D space: attributes of the triangles, or of polygons, may be queried and found everywhere inside the 2D entity. Note that attribute values may be continuous (a TIN) or discontinuous (classified polygons) within this structure.

However, if our triangulation is a TIN, with elevation values at each vertex, then our 2D graph is embedded in 3D space, and we move our F-B-BB set over to the 3D column – and also down a row, as our filled 3D field (F) – perhaps the earth below the terrain surface – is now bounded by a 2D surface (B), and it in turn by the graph of the triangle edges (BB) – and they in turn are bounded by nodes (BBB).

If instead we move our original 2D set to the left (1D space, along a curve) our filled field (F) moves up one row, and is a 1D edge bounded by 0D nodes (B). (We have no boundaries-of-boundaries.) (In 0D our space is filled with a 0D node, and that is all.) Work on 1D curves is sometimes referred to as dynamic segmentation, when attributes are assigned along portions of a curve.

We have mostly studied 2D space, with 2D fields embedded (F) – but with 1D boundaries (B) and 0D nodes (BB) (column 2). However we have sometimes worked with 1D space (embedded in 2D) for road or river networks. Our TIN model is a 2D network (on a surface) embedded in 3D, with a single z (3D) value at each 2D location. However, our 2D surface may well be more complex – it could be the exterior surface of a building, for example, where the 'field' (F) is a volume enclosed by a surface (B). We will examine some of these possibilities in the next section – but it is important to remember the distinction between the dimension of the embedding (global) space and the dimension of the structure we are building or examining. A terrain model, for example, is not a 3D structure – it is 2D. A GIS with modelled building exteriors is not 3D – its structure is 2D, as are most CAD (Computer Aided Design) models, as only the exterior is being modelled and there are no interior volume entities, such as rooms.

6.1.2 *CAD and Quad-Edges*

Earlier we looked at how to connect spatial entities together within the computer – and these same edge operators described above are not only appropriate for 2D structures (or '2.5D' with terrain elevations): they are appropriate for any '2-manifold' where the graph can be drawn: balloons, donuts, etc. In particular, they can be used to model any 'simple' solid object (without any tears in the surface, or objects only joined at a vertex or edge).

This is the basis for traditional CAD (Computer Aided Design) systems based on a 'b-rep' (boundary representation) of the object being modelled. A key aspect here is the ability to build or modify the model one edge at a time without risking invalid topological connections – the validated operations for this are called 'Euler Operators'. Probably the simplest way to construct these is by using the Quad Edge structure: valid sets of operators can be constructed entirely with 'Make-Edge' and 'Splice' as described previously. A valid set of operators is shown below: the names specify the number of edges, vertices, faces and shells (new models) added ('Made') or removed ('Killed') on the model while preserving the topology and the Euler Formula. Each operator has its inverse operation to undo its effect. Other sets of operators are possible.

It is important to demonstrate the simplicity of the method, despite its apparent complexity. The basic structure of four 'quads' and the pointer loops around nodes and faces have been described previously, along with the two operators 'MakeEdge' and 'Splice'. The very brief Pascal code is given in the 'A Little Code' box below. Splice is its own inverse, and consists of merely reassigning the appropriate loop pointers, so that two loops around the vertices and a loop around the common face become one loop around the joined vertex and two separate face loops – see the earlier figure. Repeating the operation reverts to the original condition, as face and vertex loops are duals, and equivalent. MakeEdge constructs a new isolated edge linking Pt1 and Pt2 (see Figure 72 and Figure 244, left with the appropriate minimal loops – and 'DeleteEdge' removes it from memory.

With this minimal code we are ready to examine Euler Operators, which are a set of validated operations used to modify an existing graph – the 'shell' or boundary of some polyhedron – without losing the all-important connectivity (navigability) required to predictably join portions together without losing track of the necessary neighbours. They are traditionally given names indicating the number of elements made (M) or killed (K) – thus 'MEVVFS' is a new independent shell with a new edge and two vertices, along with a new (global) face and a shell, as in Figure 72 for the Quad-Edge. 'MEV' makes a new edge and vertex, adding it to a previous graph (Figure 244, right). 'MEF' adds a new edge and face (Figure 245, splitting a previous face in two), 'SEMV' splits an edge and inserts a vertex between the two parts (Figure 246, its inverse is 'join'). Another, 'MZEV' splits a vertex in two and inserts an edge (of zero initial length) in between. Various other Euler Operators may be devised, with a minimum of five or six necessary to construct any simple (genus zero) polyhedron.) Again, the very brief Pascal code is given in the 'A Little Code' box below.

For the initial diagram on the left of Figure 246, MEVVS created the initial segment, and MEF joined the two ends and made a new face – followed by SEMV to split the edge and give a triangle.

162

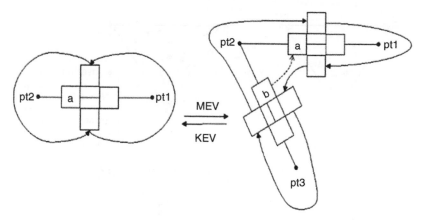

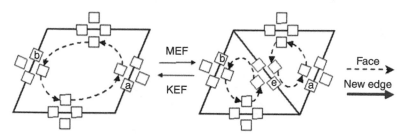

Figure 244.   MakeEdgeVertex.

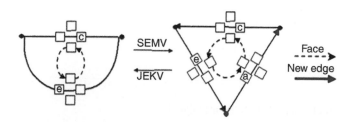

Figure 245.   MakeEdgeFace.

Figure 246.   SplitEdgeMakeVertex.

For the Delaunay triangulation algorithm, a new point pt is inserted within the circumscribing triangle by repeating MEF and SEMV as above, to give the top-left, top-right and bottom-left portions of Figure 247, and then connecting pt with an exterior vertex with MEF, as shown on the bottom right.

Following the basic algorithm, after insertion the exterior edges are tested with the InCircle predicate, and edges switched if necessary by calling KEF twice to remove the central edge, and calling MEF twice to replace it in the correct orientation (Figure 248).

A LITTLE CODE

The Quad-Edge and CAD operators are sufficiently useful, and the code itself sufficiently brief, that it is worth including the basic code at this point. The language is Delphi (Visual Object Pascal) — which is readily translated to any other pointer-based object-oriented language.

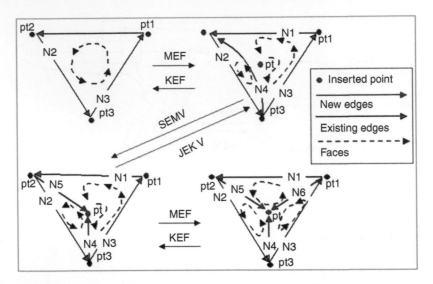

Figure 247. Adding a point in a triangle.

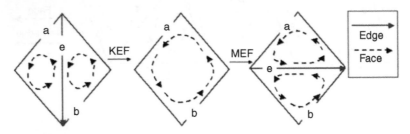

Figure 248. Switching a triangle pair.

The Quad-Edge Operators:

Make Edge
function Quad.MakeEdge (Pt1, Pt2 : Point) : Quad;
var
          Q1, Q2, Q3, Q4 : Quad;
begin
          Q1 : = Quad.Create; Q2 : = Quad.Create;
          Q3 : = Quad.Create; Q4 : = Quad.Create;
          {create four new 1/4 branches for the edges}
          Q1.R : = Q2; Q3.R : = Q3;
          Q3.R : = Q4; Q4.R : = Q1;
          {four branches link together}
          Q1.N : = Q1; Q3.N : = Q4;
          Q3.N : = Q3; Q4.N : = Q2;
{link Q1 and Q3 to themselves, Q2 and Q4 link to each other}
              Q1.Org : = Pt1; Q1.Dest : = Pt2;
              Q3.Org : = Pt2; Q3.Dest : = Pt1;
              Result : = Q1;
              {link pointer to vertices}
      end;

Delete Edge
```
procedure Quad.DeleteEdge (e);
begin
        e.R.R.R.Free;   e.R.R.Free;
        e.R.Free;       e.Free;
        {Release Memory when deleting edge}
end;
```

Splice
```
procedure Quad.Splice (a, b);
{where a, b are input Quad-Edges}
var
        Alpha, Beta, An, Bn. Aln, Ben : Quad;
begin
{get neighboring edges : Alpha and Beta in Guibas and Stolfi}
        Alpha : = A.N.R;   Beta : = B.N.R;
        An : = A.N;   Bn : = B.N;
        Aln : = Alpha.N;   Ben : = Beta.N;
        {reconnect the four pointers}
        A.N : = Bn;   B.N : = An;
        Alpha.N : = Ben;   Beta.N : = Aln;
end;
```

The simple CAD operators:
MEVVFS
```
function Euler.MEVVFS (pt1, pt2 : TVPoint) : TQuad;
{pt1 and pt2 are input parameters}
var
        a : TQuad;
begin
        a : = Quad.MakeEdge(pt1, pt2);
        result : = a;
{Make Edge function is called. Edge 'a' runs from point 'pt1' to 'pt2'}
end;
```

KEVVFS
```
procedure KEVVLS(Q1 : TQuad);
begin
        Quad.DeleteEdge(Q1);
{call function to kill edge e and release the memory}
end;
```

MEV
```
function Euler.MEV(a: TQuad; pt3: TVPoint) : TQuad;
var
        org : TVPoint; b : TQuad;
begin
        org : = a.Org;
        b : = Quad.MakeEdge (org, pt3);
        Quad.Splice(a, b);
        result : = b;
{MakeEdge and Splice are called. Edge 'b' is created which runs from point 'pt2' to 'pt3'}
end;
```

KEV
```
procedure Euler.KEV (a: TQuad);
var
          b : TQuad;    dest : TVPoint;
begin
          b : = a.Oprev;
          Splice(b, a);
          DeleteEdge(a);
{disconnect the input edge 'a' from with the next edge by using Splice in Quad-Edge}
end;
```

MEF
```
function Euler.MEF(a, b : TQuad) : TQuad;
var
          e : TQuad;
          pt3, pt1 : TVPoint;
begin

          pt3 : = a.Org;
          pt1 : = b.Org;
          e : = Quad.MakeEdge(pt3, pt1);
          Splice(a, e);    Splice(e.Sym, b);
          {Edge 'e' is made and connects to edges a and b}
          result : = e;

end;
```

KEF
```
procedure Euler.KEF(e : TQuad);
var
          a, b : TQuad;
begin
          a : = e.Oprev;    b : = e.Lnext;
          Splice(a, e);      Splice(e.Sym, b);
          {disconnect edge 'e' with 'a & b' edges}
          DeleteEdge(e);
          {release memory}
end;
```

SEMV
```
function SEMV(e : TQuad; pt : TVPoint) : TQuad;
var
          pt1, pt2 : TVPoint;
          a, c : TQuad;
begin
          pt1 : = e.Org;    pt2 : = e.Dest;
          c : = e.Lnext;
          Splice(e.Sym, c);
          {disconnect the input edge with its neighbours}
          e.Dest : = pt;
          {change the destination of the edge 'e'}
          a : = Euler.MEF(e.Sym, c);
          {MEF is used to make a new edge and face}
          result : = a;
end;
```

JEKV

```
procedure JEKV(e : TQuad);
var
        a, c : TQuad;
begin
        a : = e.Lnext;
        c : = a.Lnext;
        Splice(e.Sym, a);
        Splice(a.Sym, c);
        {disconnect the edge 'a' with its neighbour edges}
        e.Dest : = a.Dest;
        {change the destination of edge 'e'}
        Splice(e.Sym, c);
        a.Dest.Free;
        Quad.DeleteEdge(a);
        {delete vertex 'pt' and edge 'a'}
end;
```

6.1.3 *TIN models and building extrusion*

This gives us our terrain model. If we wish to extrude a building from this terrain we can use these operations to 'build' the vertical walls: for a simple example we will just imagine a triangular building. One sequence of operations could be as shown: first use MEV to make a wall face (Figure 249); then use SEMV twice to give the wall's roof points (Figure 250); then use MEV and SEMV to make the second wall (Figure 251); then use MEV to form the third wall and the roof (Figure 252).

Note that, while the interior edges appear to be on the same plane as the exterior triangle in the figures below, their elevation values lift them up to the roof level. In addition, their (X, Y) coordinates are the same as those for the base triangle – giving vertical walls. (Note: this is impossible for a Delaunay triangulation, showing that this kind of modelling does not conform to our previous DT/VD work.)

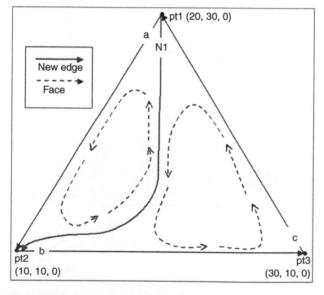

Figure 249.   Making a wall face.

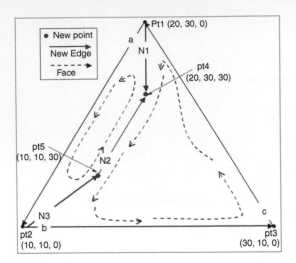

Figure 250.   The wall's roof points.

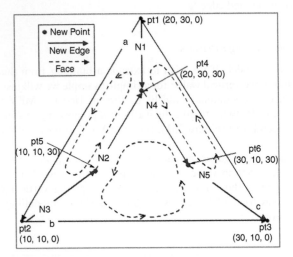

Figure 251.   Making the second wall.

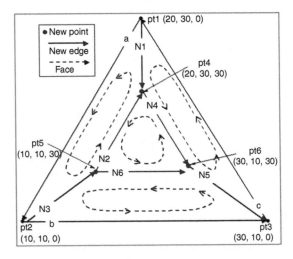

Figure 252.   Finishing the third wall and the roof.

6.1.4 *Bridges and tunnels*

This procedure is sufficient to model plausible buildings extruded from the landscape: so far we are merely rephrasing CAD Euler Operators in Quad-Edge terms – but simplifying the process. (In the original work by Mäntylä the half-edge structure was used, but the Quad-Edge is simpler.)

Now we wish to go beyond our simple TIN modelling – even with extrusions – and look at models with bridges and tunnels ('handles' in topology terminology), using the same operators – this is possible because Euler Operators are valid on any 2-manifold.

We use MEHKF/KEHMF (Make Edge and Hole, Kill Face – or Kill Edge and Hole, Make Face). Its coding is exactly the same as for MEF/KEF, but whereas MEF assumes that the two vertices being joined are on the same face loop, MEHKF assumes that they are not – giving an edge between two unconnected faces, and converting the two faces into a 'hole' plus a face forming a 'pipe' through the model. (Think of a sheet of paper curled over so two opposite edges meet. In both diagrams we are looking at the triangles from 'inside' the surface. Exactly the same process may construct a 'bridge' instead.) Figure 253 shows the initial TIN and the selected triangles.

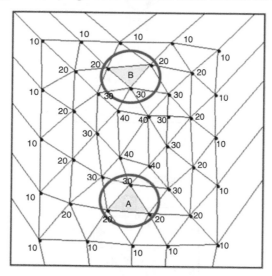

Figure 253.   Select two triangles in the TIN.

Figure 254 shows the initial pair of triangles for the tunnel (Figure 254: we are looking from the inside).

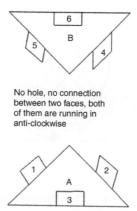

Figure 254.   Interior view.

169

Figure 255 shows making a hole/bridge using 'MEHKF': the face loop follows around triangle A, then along one side of the new edge, then around B and back to A along the other side of the new edge (Figure 255). This one curled-up face may be split up in the usual way using MEF, to give a tunnel you can walk through!

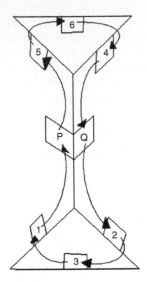

Figure 255.   Tunnel face joining the triangles.

Reshape the tunnel wall with MEF (Figure 256, Figure 257).

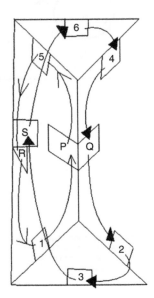

Figure 256.   Split the face.

The resulting hole is like Figure 258: we are looking from 'inside' the solid of the polyhedron, the 'rock', so loops appear to be running clockwise.

Finally in Figure 259 we split the rectangular faces into triangles, to conform to our TIN models.

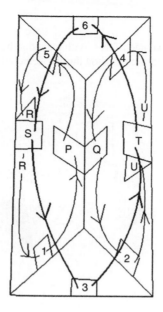

Figure 257.   Split the face again.

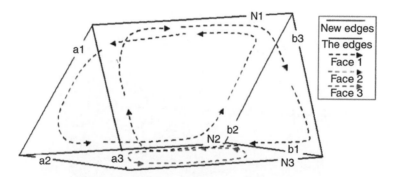

Figure 258.   Completed tunnel.

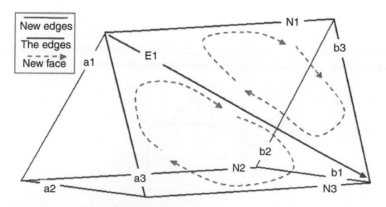

Figure 259.   Triangulating tunnel walls.

6.1.5   *The Extended TIN*

Here are some examples of the Extended TIN – with extruded buildings and 'handles'.

Figure 260 is a simple TIN with a hole, showing the selected triangles. Figure 261 shows the tunnel from above.

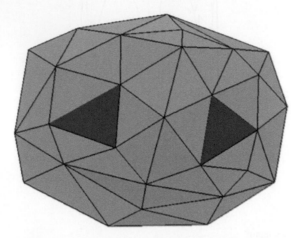

Figure 260.   Selected triangles.

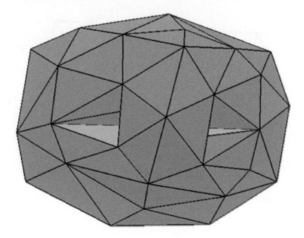

Figure 261.   Tunnel from above.

Figure 262 shows the tunnel from the side. Figure 263 shows the initial triangular hole enlarged using the same 'Swap' operator used for the Delaunay triangulation.

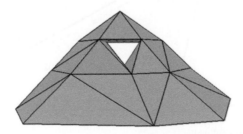

Figure 262.   Tunnel from the side.

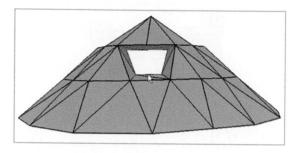

Figure 263.   An enlarged tunnel.

Figure 264 gives a rather simple extruded triangular building. Figure 265 adds a second building, and Figure 266 shows them connected by a bridge.

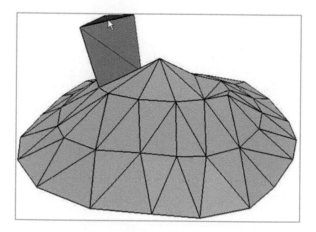

Figure 264.   A simple extruded building.

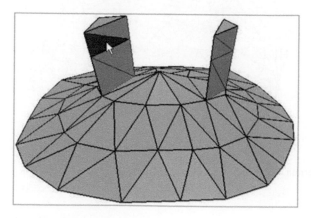

Figure 265.   A second building.

Note that bridges and tunnels are topologically identical: in one case we are looking from 'outside' and in the other we look from 'inside'.

On another terrain model (Figure 267) two bridges are constructed (Figure 268).

173

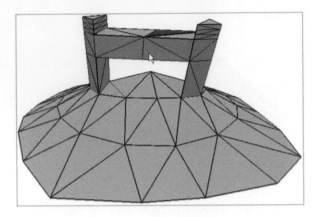

Figure 266.   Adding a connecting bridge.

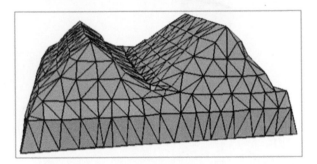

Figure 267.   Another terrain model.

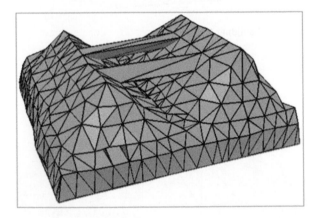

Figure 268.   The terrain model with two bridges.

Finally, Figure 269 shows a more complex building.

This concludes our discussion of terrain modelling with TINs and their extensions. We have looked at the basic concepts of triangulations, planar graphs and their duals, the importance of the Voronoi/Delaunay dual relationships, and the necessary arithmetic predicates and coordinate systems, and we have outlined the incremental Voronoi/Delaunay algorithm for TIN modelling.

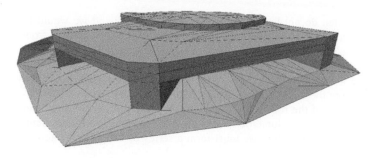

Figure 269. A more complex building.

So we have shown that TIN modelling is just a subset of CAD, and the two can be considered to be the same.

FURTHER READING

Much has been said in the mathematical literature about dimensionality, but Gold (2005) gave a useful summary for GIS: topographic surfaces are 2D surfaces embedded within 3D space, for example, and their data structures are therefore 2D – and not 3D, nor even really 2.5, or '2.75' as in the work of Tse (2002, 2003, 2004) with bridges and tunnels.

Mäntylä (1986, 1988) first defined what are now called 'Euler operators' – operations that, when performed on valid inputs, are guaranteed to result in a valid 2D surface obeying the Euler-Poincaré formula. Lee (1999) wrote an excellent textbook on Computer Aided Design (CAD) with a clear outline of the required topology and data structure issues.

6.1.6   *LiDAR building models: the walls*

On this basis, we can combine a terrain model with building outlines, to produce a simple city block model (Figure 270).

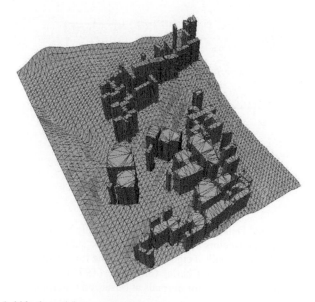

Figure 270.   Extruded block models.

However, we often do not have the building ground plans, and we would like to have more realistic roofs – in this case we need to obtain LiDAR data and reconstruct from that. This takes several logical stages, especially if we are not working from predefined building templates, which are often used where buildings are of similar forms. Firstly we have to extract the wall boundaries (which we will assume correspond to the outlines of the eaves). Secondly we must extract the roof planes from the observations falling within the building outlines. Thirdly we must calculate the intersections of the roof planes, and extrude the building exteriors, and fourthly, where possible, convert this to a volumetric model and add interiors from other sources. As we have no pre-defined building structure we must form a series of Propositions about the forms of the buildings.

Proposition 1: Buildings are collections of contiguous elevations that are higher than the surrounding terrain. Their boundaries are 'walls'.

There are various ways we might attempt to find these boundaries, but in our approach we apply a coarse Voronoi diagram over the original data, with perhaps 50–100 LiDAR points in each cell. We then attempt to modify these cells so that the building boundaries (defined as a partition between 'high' points and 'low' points) are a subset of the Voronoi cell edges.

In Figure 271 we divide the region into sample cells.

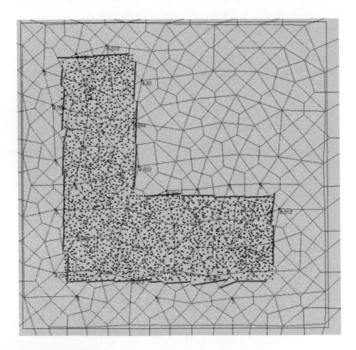

Figure 271.   Sampling cells for LiDAR data.

We then take all the original Delaunay triangles in each sample Voronoi cell, look at their vector normals, and calculate the eigenvectors and eigenvalues of the sum-of-squares matrix.

The eigenvector which has the smallest eigenvalue gives the line of sight (fold axis, Figure 272). This is based on the geological work of Charlesworth et al. (1975), who used strike-and-dip field measurements of outcrops to obtain vector normals at observed outcrops of the folded strata they wished to model.

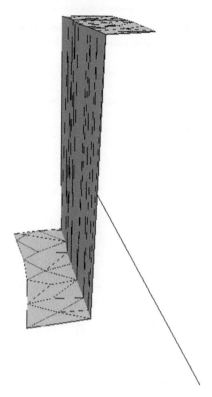

Figure 272.   Eigenvector with the smallest eigenvalue.

Proposition 2: Walls have a specified minimum height, and this height difference is achieved within a very few 'pixels'.

The next step is to locate the line, parallel to this smallest eigenvector, that best separates 'high' elevations from 'low' ones within each Voronoi cell. For each parallel slice we use a T-test to see which position gives the largest separation (Figure 273).

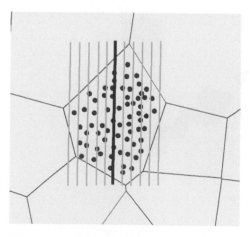

Figure 273.   Best separating line.

The Voronoi cells are then split along these lines, by adding a generator on each side, at the mid-point. This gives a set of 'high' Voronoi cells surrounded by 'low' ones.

Proposition 3: A building consists of a high region entirely surrounded by walls.

This must form a closed region, or else the high region is not considered to be a building (Figure 274).

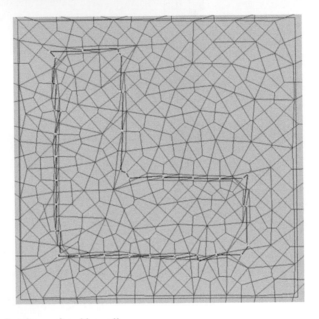

Figure 274.   Closed region enclosed by walls.

Figure 275 and Figure 276 show a block model of the Voronoi cells before and after the split.

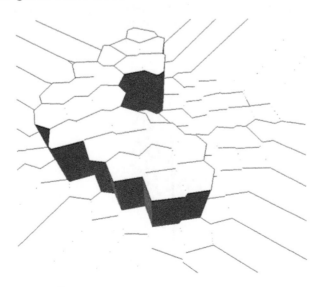

Figure 275.   Initial Voronoi cells.

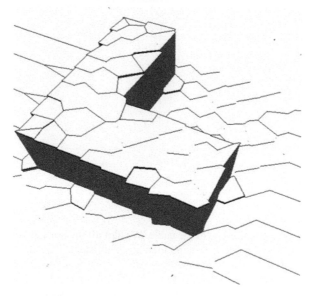

Figure 276.   Voronoi cells after split.

Other techniques can be used to detect building walls – for example, find the roof planes first and then extrapolate the walls.

6.1.7   LiDAR building models: simple roof modelling

It would be a good idea to add roofs to our block model. We will assume roofs are constructed of planar sections:

Proposition 4: Roofs are made up of planar segments, most of whose constituent triangles have similar vector normals.

Our objective is to cluster triangles with similar orientations. Each triangle has its own normal vector.

We extract the normal vectors (Figure 277) and calculate the eigenvalue of the variance-covariance matrix, as before. The eigenvector with the smallest eigenvalue gives us the line of sight (Figure 278, the fold axis) – the same method we used for wall detection. This method only works with a simple gable roof.

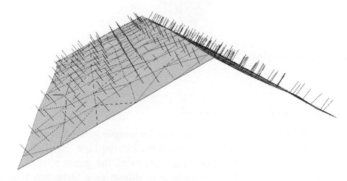

Figure 277.   Vector normals of roof triangles.

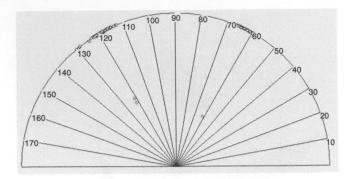

Figure 278.   Vector clustering along the fold axis.

This gives us a 'view' along the axis of the gable. We plot all the normal vectors on a semi-circle perpendicular to the smallest eigenvector, and look at the clusters. This identifies the number and orientation of the groups of similarly-oriented triangles – presumably representing individual roof planes.

A plane may be defined by its vector normal plus a point on its surface. First of all we average the orientations of each triangle in the cluster. Then we project the vertices of these triangles onto this vector normal, and use the average as the reference point on the roof plane. This gives us the equation of the plane, which can be intersected with the other roof planes and walls to give a simple roof, rather than just a block model (Figure 279).

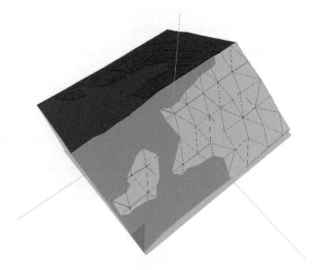

Figure 279.   Roof plane, and original TIN.

6.1.8   *LiDAR building models: compound roofs*

The eigenvector method, however, is only appropriate for simple gable roofs; for more complicated structures we need to analyse the vector normals on the hemisphere, rather than the semicircle – adding another dimension. Figure 280 shows the vectors of the gable roof plotted on the hemisphere – two clusters are evident. (The hemisphere is shown as a Delaunay triangulation of the projected vector normals.)

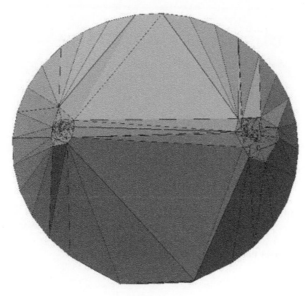

Figure 280.   Vector normals plotted on the hemisphere.

Looking at a more complex structure – an L-shaped building – we have the following triangle distribution (Figure 281).

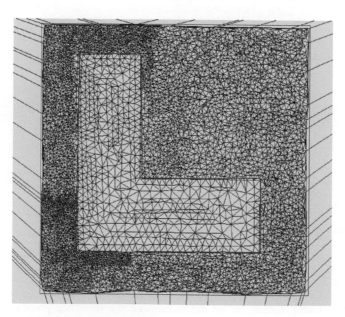

Figure 281.   TIN model of complex building.

Figure 282 gives the following projection onto the hemisphere – four clusters can be seen.

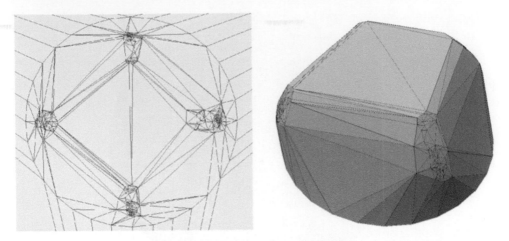

Figure 282. DT of the Vector normals, projected on the unit hemisphere.

To identify the clusters, we use the Minimum Spanning Tree of the DT, described previously. This gives a tree with four close sub-trees, with wide gaps between them – we cut the branches if an edge is longer than five degrees of arc (Figure 283).

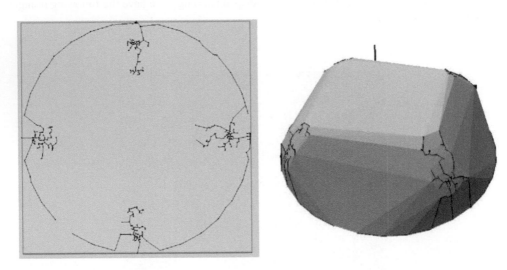

Figure 283. The MST of the DT.

However, our four clusters may contain vector normals from geographically separated roof planes – roof ends as well as mid parts in our example. These may be separated by projecting onto the average vector normal, as we did to estimate the reference point on the roof plane. (They will fall into two distinct clusters, which can be separated.) Figure 284 shows the resulting roof planes.

Each intersection point of three roof planes or walls may be calculated from their vector normal equations, and each plane clipped by the others. This gives an extruded model plus a roof, suitable to be inserted into the TIN as the basis of a city model – but with the buildings 'glued' to the ground.

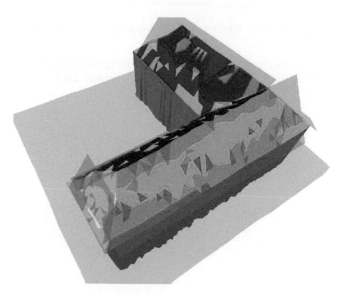

Figure 284.   Roof planes of the complex building.

The polygon set, as with Voronoi cells, has a dual triangulation.

Proposition 5: The relationships between roof planes may be represented as a dual graph.

Figure 285 shows the dual relationships between roof faces and edges – and their action as the air/earth interface.

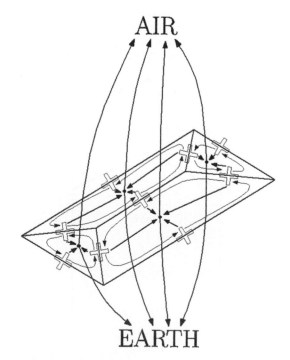

Figure 285.   Dual roof faces and edges.

Proposition 6: Building exteriors, together with the adjacent terrain, form a portion of the global 'Polyhedral Earth'.

As an extrusion, our building model is part of the underlying terrain, and is part of a CAD-type shell of the whole globe – with air above and rock beneath (Figure 286).

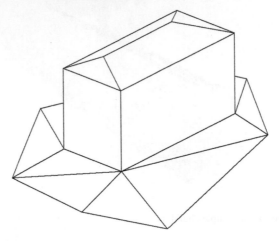

Figure 286.   Building extrusion.

If sufficient data is available, interior features may be introduced, resulting in a true 3D model, with multiple volumetric elements – the subject of the next section.

What we have done here is extended/extruded the complex polyhedron that forms the 'Earth'. It is a single polyhedron because of the unifying effect of gravity (ignoring oceans and the atmosphere). It may have many holes/bridges in it, but it is one single polyhedron, of high genus.

What we have not done is describe various volumetric objects within 3D space – just 'inside' and 'outside'. To go further we need to add volumetric entities to our current list of face, edge and node entities used in 2D GIS, or building exteriors – to give us 'True 3D' (Figure 287). Perhaps the easiest way to start is to extend our 2D point VD/DT to 3D, and then look at buildings later.

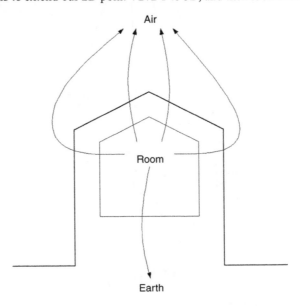

Figure 287.   Interior volumes.

FURTHER READING

Tse et al. (2007, 2008a, 2008b) examined the problem of extracting building shapes from raw LiDAR data, based on two premises: vector normals of the triangulated data points cluster to indicate the individual building planes – see Charlesworth et al. (1975) – and the building form, or template, is not known beforehand – the data-driven approach. Maas and Vosselman (1999) exemplify the model-driven approach.

6.2   SOLID 3D: REAL VOLUMES

6.2.1   *3D VD/DT*

The VD/DT is constructed from a set of point generators with a metric dominance test, using the circumcircle in 2D and the circumsphere in 3D (Figure 288). Thus the spatial relationships are constructed automatically, unlike a 2D map or a 3D building.

The InSphere test is the same determinant used in the InCircle test, with the addition of another vertex (Point 4) and another dimension (Z). On this basis we can use the incremental algorithm: walk through the tetrahedral mesh using the CCW test (again with an extra vertex and dimension) to add one vertex at a time; split the resulting enclosing tetrahedron into four; and test each exterior face to see if it is 'Delaunay'. If it is not, then interior and exterior tetrahedron pairs need to be 'flipped', the new exterior tetrahedron pairs put on a stack for later processing, and the process continued until the stack is empty.

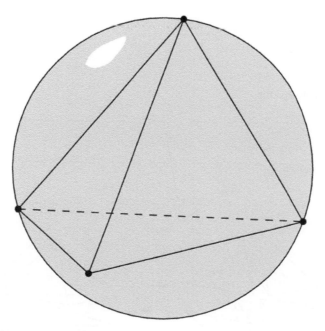

Figure 288.   Circumsphere of a tetrahedron.

However, in 3D the 'flip' process is considerably more complex: for further details see Ledoux and Gold (2007, 2008). Figure 289, Figure 290 and Figure 291 show the possible combinations.

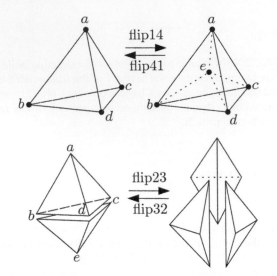

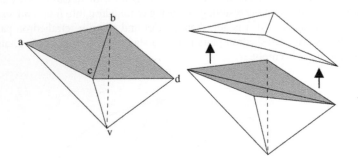

Figure 289.   1-4 and 2-3 tetrahedron flips.

Figure 290.   2-2 tetrahedron flip.

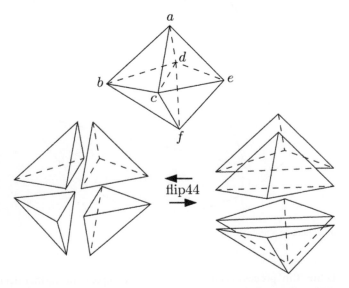

Figure 291.   4-4 tetrahedron flip.

When completed we have the 3D DT (on the left), or VD (on the right of Figure 292).

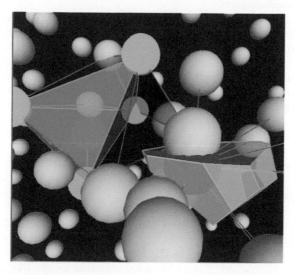

Figure 292.   3D Delaunay and Voronoi cells.

Given the 3D DT, contour surfaces may be generated by slicing the tetrahedra with the appropriate contour values. Here oceanographic temperature data is being examined (Figure 293).

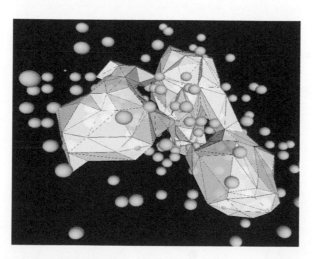

Figure 293.   3D contours from Delaunay tetrahedra.

The 3D VD is particularly useful when processing highly anisotropic data such as ships' soundings – bathymetric depth readings, often taken regularly along the ship's course.

In addition, views of any cross-section of the ocean model may be generated by simple linear interpolation within each tetrahedron – using exactly the same approach as in 2D. Here we are not projecting the data points onto the 2D plane, but are using the plane to generate the grid whose vertices are used as query points for 3D interpolation. See Ledoux and Gold (2004).

The 3D VD may also be used for proximity analysis – determining the neighbouring points to any selected point. As in 2D, the Minimum Spanning Tree of the data set is a subset of the edges of the Delaunay tetrahedralization.

It is also possible to develop the algorithm for point deletion in 3D, thus allowing Sibson interpolation, giving relatively smooth estimations – either of a cross-sectional grid or within a full 3D grid. This would also permit 3D map editing by the insertion and deletion of individual data points.

The implementation of point movement in the 3D VD/DT, and maintenance the neighbourhood relations during movement, provides the opportunity to simulate 3D flow, as in marine pollution plumes or atmospheric modelling. The techniques are very similar to the 2D case.

These techniques primarily use the 3D VD or the 3D DT – not so much both together. However, the full 3D context – VD plus DT – may be required for some applications. Sasanov et al. (2007) used them for electro-magnetic 'co-volume' integration schemes to implement Maxwell's equations Maxwell (1864) for the relationships between electric current and the magnetic field, which are based on the interpenetration of the primal and dual structures.

FURTHER READING

Much work has been done in Computational Geometry on the construction of the point VD/DT in 3D Euclidean space: it is a difficult problem due to various degenerate cases and the limitations of computer arithmetic. Liu and Snoeyink (2005) provide a useful comparison for the DT, and Ledoux (2007) gives a relatively straightforward description for the VD, while Sugihara and Hiroshi (1995) described the difficulties. Ledoux and Gold (2007) describe a structure – the Augmented Quad-Edge – for storing both the VD and DT simultaneously, and Ledoux and Gold (2004, 2008) give examples of applications in oceanography and geology. Sasanov et al. (2007) give a very interesting example of the combined VD/DT for the analysis of electrical and magnetic fields, following Maxwell (1984) – by their nature these require 3D primal/dual analysis.

6.2.2 *The augmented quad-edge and the dual half-edge*

The result of either the 3D VD or DT construction is a set of contiguous space-filling cells – a cell complex (Figure 294).

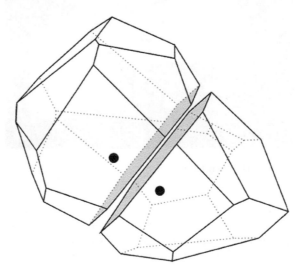

Figure 294.   3D VD cells.

However, the data structures we have been working with only operate on a single cell boundary – a b-rep model: we can navigate anywhere around a single volumetric cell, but we cannot navigate from cell to cell. This can be resolved by looking at full 3D Poincaré duality (Figure 295). In 2D the dual of a vertex is a face (and the dual of a face is a vertex) and the dual of an edge is a 'penetrating' edge (e.g. a Voronoi edge is penetrated by a Delaunay edge, and vice versa). In 3D the dual of a vertex is a cell (and vice versa) and the dual of an edge is a face (and vice versa). Thus all our 3D structures may be represented by vertices and edges – the components of a graph. A Voronoi cell may be represented by its Delaunay vertex and a Delaunay cell (tetrahedron) is represented by its Voronoi vertex. A Voronoi cell face is represented by its penetrating Delaunay edge and a Delaunay tetrahedron face is represented by a penetrating Voronoi edge.

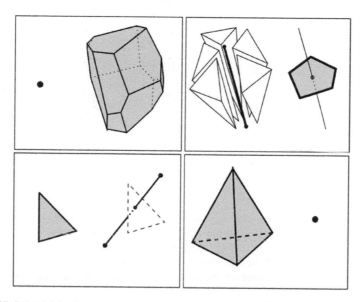

Figure 295.   3D Poincaré duality.

The remaining question – how to move from cell to cell – is thus resolved: adjacent cells have a common face, and thus have a common penetrating dual edge. So going to the dual edge between two cells permits navigation from cell to cell: whatever our particular focus of interest (the VD or DT at the moment), the dual structure is required to permit movement between the neighbouring cells. The question is thus: how to link the two graphs so that the required edge may readily be found? A new data structure is required.

In the 3D volumetric case a surface-graph based data structure is inadequate. Alternative data structures include sets of tetrahedra, and CAD-type radial-edge or equivalent data structures, although these are difficult to manipulate.

Let us use our previous PAN graph analysis, and the Poincaré duality, to derive possible 3D structures (Figure 296). Here F represents faces (or polygons), E represents edges (or arcs) and N represents nodes. A new element, V for volume entities, is introduced.

In Figure 296, (a) represents a traditional 2D triangle data structure: each triangle (a face) has pointers to the three adjacent faces, and to the three vertices. Edges are not represented. Nodes have no pointers.

Figure (b) shows an edge-based representation of a triangulation. Each triangle edge has pointers to two (perhaps anticlockwise) adjacent edges – or else all four (clockwise plus anticlockwise). An edge has pointers to the two nodes at the ends and the two faces on its opposite sides. (This is essentially the same as the Winged-Edge structure.)

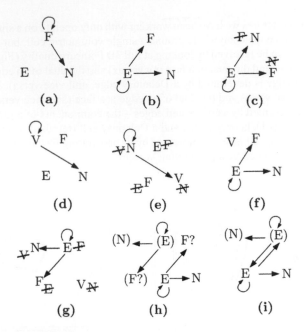

Figure 296. PAN graph combinations.

Figure (c) shows the dual of (b) – e.g. the Voronoi diagram rather than the Delaunay triangulation. All that has changed is each (Delaunay) face has become a (Voronoi) node, each (Delaunay) node has become a (Voronoi) face, and the initial Delaunay edge has become its matching Voronoi edge.

If we move to 3D, we would have a Delaunay tetrahedralization – Figure (d) – with volumes (V) having pointers to the four adjacent tetrahedra and to the four vertices. Edges and faces are not represented.

Figure (e) shows the dual of (d). The tetrahedral volume has become a Voronoi node, and the tetrahedral node has become the associated Voronoi volume. Faces and edges are not used directly, but the Delaunay faces are replaced by their penetrating Voronoi edges, and the Delaunay edges have become the Voronoi faces that they penetrate.

Figure (f) shows an edge-based representation of a tetrahedralization. Each edge has pointers to all adjacent edges, connected at its vertices and permitting navigation, as well as pointers to the four nodes and the four faces: volumes are not yet incorporated.

Figure (g) shows the dual of (f). Edges connecting (Delaunay) nodes are replaced by the matching penetrated faces in the (Voronoi) model; Delaunay faces are replaced by the penetrating Voronoi edges, and (Voronoi) edges have pointers to other Voronoi edges to permit navigation – this is exactly as in 2D in Figure (c). Delaunay volumes become Voronoi nodes, and Delaunay nodes become the generators of Voronoi volumes.

Putting (f) and (g) together we get Figure (h), with the elements from the dual put in parentheses. 'F?' indicates that the face element may be eliminated and replaced by the penetrating edge. This gives us the possibility of a pure graph representation, composed entirely of nodes and edges, with face information attached to the appropriate edge. The same is true for volume elements: their attributes are associated with the dual nodes.

Figure (i) gives the result when the two edge-based graph structures, one in the primal and one in the dual, are merged: in each case the face entity is replaced by the edge entity in the dual. We thus need some kind of pointer between primal and dual graphs. As this is an edge graph, with all navigation pointers in the edges, and the individual b-reps are modelled using the Quad-Edge, it is appropriate to add a pointer between the quad associated with a particular edge-face pair (qf) and the quad associated with the edge-face pair (qf.through) of the dual edge penetrating the original face, as in Figure 297.

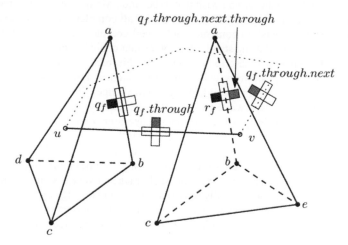

Figure 297. The 'through' pointer.

However, there are several (n) edges on the original face, and thus n edges pointing to it in the dual. Each of these dual edges is part of the cell surrounding a vertex of the original face. Thus if u and v are the volume nodes in the dual, there will be n edges connecting them, one for each of the volumes surrounding the n vertices of the original face (three in Figure 297 – nodes a, b and c). For instance, q_f is associated with node a, because q.vertex = a and q.rot = q_f. Node a is enclosed by a dual volume, which includes q_f.through. Two other edges, bounded by dual nodes u and v, are parts of two other volumes enclosing nodes b and c.

This gives us a navigation structure – the Augmented Quad-Edge (Figure 298) – that permits pointer-based movement around each original b-rep shell, and then to a penetrating dual edge, and then, after switching to the other end of this edge using 'sym', back to the primal cell complex – but this time onto the adjacent cell to the original one. Full navigation is then possible – except at the boundary of the whole cell complex. This is handled by creating one large enveloping cell, 'inside-out', which does the same job as the 'exterior' face of a 2D map.

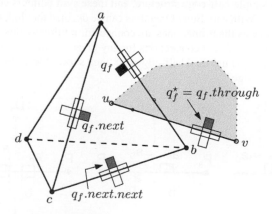

Figure 298. 'Through' pointer connections.

Full pointer-based navigation is thus defined. This is important, as in 2D, as it is the basis of all local queries of the map – starting at any edge (often the previous one used) we can 'walk', edge by edge, to some desired destination edge, and from there examine the adjacent nodes, volumes, faces and edges in both the primal and the dual graph. ('Primal' and 'dual' are interchangeable: we start with the primal, whichever you wish that to be, and switch to the dual as required – this is then repeated to revert to the original.) This will hold for all complete cell complexes.

Problems arise when we wish to construct such a cell complex. This was possible for the 3D VD/DT model as this was built using standard elements that could be snapped together in standard ways, using the 'flip' operators described previously: this method was difficult, but possible. However, when constructing arbitrary cell complexes, such as buildings with rooms, no such standard elements exist, apart from individual edges and vertices. Edges must therefore be added one at a time to the 'primal' graph while ensuring that the 'dual' graph is updated simultaneously.

In CAD (Computer Aided Design) a set of standard operators have been developed for straight-forward b-rep models that guarantee that connections will be complete if the input parameters (specified edges) are valid. These 'Euler Operators' work on simple 2-manifolds, with or without holes, and there is no need for a 3D dual structure as the only volume elements are 'inside' and 'out-side'. However, the 2D dual, on the surface of the shell, makes it very easy to construct them: they may be made with one or two (occasionally three) calls to the Quad-Edge operators MakeEdge and Splice. Thus the Quad-Edge operators form the lowest level that preserves the connectedness of a navigable graph and the Euler Operators add the guarantee that the surface (shell) being constructed is valid and continuous.

The Augmented Quad-Edge then adds a link between the graph of each primal cell and the graph of an associated dual cell – thus permitting navigation between adjacent cells in the cell complex.

We needed to find a 3D equivalent to the 2D Euler Operators, where individual edges could be added to an existing navigable graph, and at the completion of each operation the graph was still navigable – that is, unbroken – and the surface of each shell remained valid. In addition, the same properties must hold for the associated dual graph. We found no way of doing this while using the underlying Quad-Edge that we used for the earlier work.

The Quad-Edge was formed by joining the primal graph to the dual graph by using the 'Rot' pointer to cycle through the 'Next' pointers: first in the primal, then in the dual, etc. (Figure 72). We could have subdivided this into two parts: the primal (linking the two half-edges together as in the traditional half-edge data structure) and the dual (likewise) – but then the primal and dual graphs would not be linked at all. Alternatively, we could permanently associate each primal half-edge with its matching dual half-edge, and allow the half-edge in each space to be disconnected from its matching half – and be reconnected as required in the construction process. (For example, link Q1 and Q4 together in Figure 72 and Q2 and Q3.) Initially Q1–Q4 would have a sym pointer to Q2–Q3, and vice versa, as in a simple half-edge structure, and these sym pointers could be re-assigned in the construction process. With care Euler Operators can be defined that link the dual parts together correctly at the same time as the primal ones are connected: intuitively this should be possible, as there is only one correct set of dual connections for any given connection operation in the primal graph. Figure 299 shows a primal-dual half-edge pair, then shows two of them snapped together – forming the usual Quad-Edge.

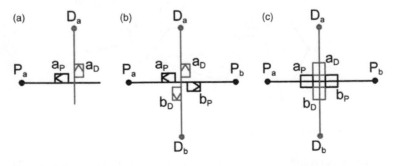

Figure 299.   Primal-dual half-edge pairs.

192

Now let us re-examine the idea of Poincaré duality. In 2D the dual of a 1D edge is a penetrating 1D edge; in 3D the dual of a 2D face is a 1D penetrating edge. Thus in 2D the dual edge will penetrate into (in the same plane as) the polygon face; in 3D the dual edge will penetrate (perpendicular to) the face of the polyhedron. Thus the links between the primal and dual edges will have a different order.

In 2D if we go to the dual of edge e (q_f.rot), go to its neighbour (q_f.rot.sym), and look at its dual (q_f.rot.sym.rot) we will end up at the original e.

In 3D if we go to the dual of edge e (q_f.through), go to its neighbour (q_f.through.sym), and look at its dual (q_f.through.sym.through), we end up on an edge pointing at the same vertex but located on the adjacent face of the neighbour polyhedron.

The 3D dual model does not include the 2D dual – it does not include an edge connecting the 'centres' of the two adjacent faces of e.

It can be shown that we can 'snap' half-edges together in this 3D model just as we can in 2D: because primal and dual are permanently linked we have sufficient information, once we have connected our primal half-edges, to know how to connect their associated duals. Thus in building a model of, e.g. a house, we can connect the half-edges forming the corners of rooms and have the duals formed automatically.

In the Augmented Quad-Edge (AQE) structure individual b-reps (cells) are modelled using the Quad-Edge, so it is appropriate to add a pointer between the quad 'q_f' associated with a particular edge-face pair and a quad associated with the edge-face pair of the dual edge penetrating the original face. In other words, part of a primal face loop links with part of the bundle of dual edges penetrating the face. This pointer was labelled 'Through' in the AQE. 'Adjacent', the quad 'back-to-back' with q_f is thus q_f.through.next.through.rot^2. ('q_f' was labelled 'q' in the original article, and in Figure 300.)

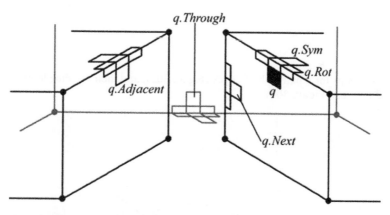

Figure 300. The Augmented Quad-Edge.

The AQE was used to model well-defined cell structures, such as the 3D VD/DT, where the only operations were the various 'Flip' operators – these could be painstakingly sewn together, converting between the 'before' and 'after' conditions. However, it was inappropriate for the construction of arbitrary shells where 3D 'Euler Operators', similar to 2D Euler Operators, were required. This required the splitting of the four parts of the Quad-Edge, as described above.

The Dual Half Edge (DHE) builds on the AQE. Consecutive quads (Q2–Q3 and Q4–Q1 in the Figure 72, or q_f-q_f.rot and q_f.rot^2-qf.rot^3 in Figure 301) are merged and represented as half-edges e and e.S respectively. The DHE 'quad' therefore required pointers to the vertex (V), the second half of the edge (S), the linked edge in the dual (D, just described), NV (the next edge around the vertex in the shell) and NF (the next edge around the face associated with the original edge). (This last is not required for adjacent cells connected by faces or edges, but is required to handle cells connected only by a vertex.)

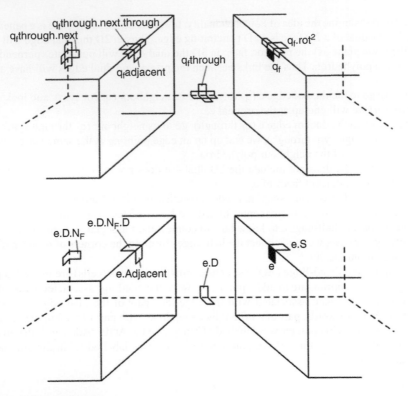

Figure 301.   The Dual Half-Edge.

Thus the N_V loops for a pair of half-edges (Figure 302), and their N_F face loops (Figure 303), and the Dual pointers (Figure 304).

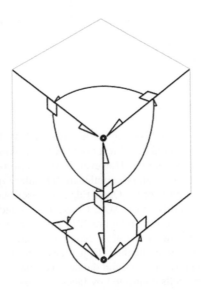

Figure 302.   Vertex loops.

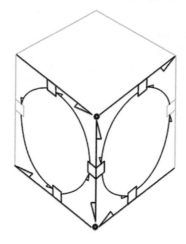

Figure 303.   Face loops

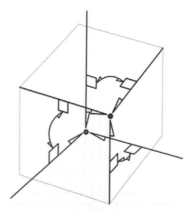

Figure 304.   Dual pointers.

Thus, in Figure 305, the following pointers and navigation operations are:

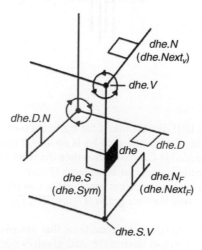

Figure 305.   Dual Half-Edge pointer structure.

Pointers
N_V (next CCW* edge around a vertex)
N_F (next CCW* edge around a face)
S (the second half of an edge)
D (dual edge)
V (vertex)

Navigation:
Sym (he.S)
$Next_F$ (he.N_F)
$Next_V$ (he.N_V)
Dual (he.D)
Adjacent (he.D.N_F.D.S)
$Next_E$ (he. D.N_F.D)

*we look at a cell from outside

Because each face has 'n' edges there will be n linked edges penetrating that face, forming a 'bundle' of edges, as shown in Figure 306. This appears wasteful of storage, but all non-manifold data structures have similar requirements.

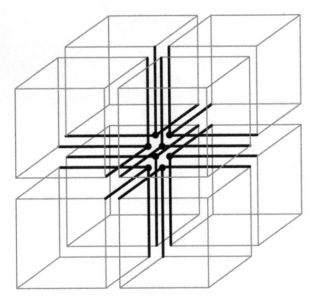

Figure 306.   Bundles of edges.

A simplified version of the DHE is possible. If the connection of shells by a vertex alone is not possible in the application being envisioned, then N_F is no longer needed. If dual vertices (volume elements, or rooms) are not needed in the application then the whole dual graph may be eliminated and the D pointer used to directly connect adjacent faces (Figure 307).

This makes graph traversal between adjacent rooms more complex as it is no longer possible to navigate from room to room using the dual graph (Figure 308, Figure 309) – we have to go from edge to edge.

The DHE thus constructed satisfies some conditions that are easy to describe but difficult to implement. Unlike other non-manifold models the dual graph is constructed simultaneously with the primal graph, and achieved automatically, based on the underlying principle that, like the 2D

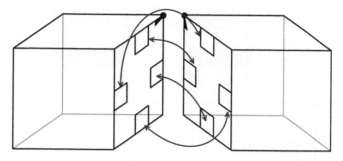

Figure 307. Simplified DHE.

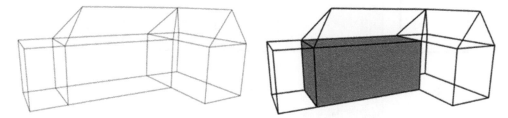

Figure 308. A 'room' volume element.

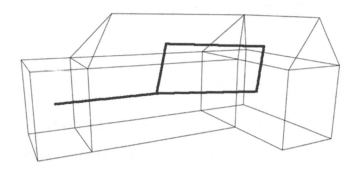

Figure 309. Dual adjacency graph.

Quad-Edge, the partially-constructed graph is fully navigable after the completion of each Euler Operator.

Construction of a single cell (without the dual) is a simple process using traditional Euler operators. For non-manifold models it is required to be able to construct more complex structures: to create non-manifold cell complexes the standard Euler operators should be extended to manage connections between cells and to include operations like joining two cells by a shared face, edge or vertex. However, cells joined by a face is the most common situation in cell complex construction and building models.

Thus we have a real-time locally-modifiable structure that integrates the primal and dual graphs – as well as their attributes, which may be attached to any vertex, half-edge, face or volume entity – as these last two are stored as dual edges and nodes. This fits the original objective, which was building interior modelling for path planning – even under changing conditions, such as locked doors.

Attributes can be assigned to primal nodes and edges as well as to dual elements. Thus in Figure 310 half-edge 1985 penetrates from the central node 15, through the green wall to node 10. Further attributes are the connection weight of 35 (perhaps distance), the existence of a door, and

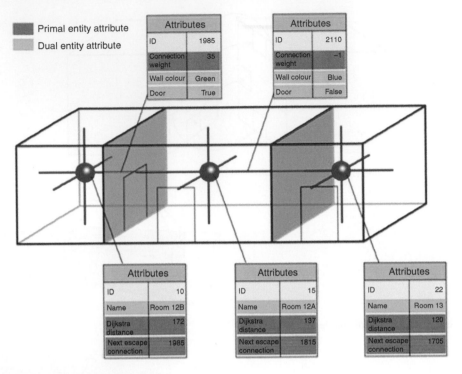

Figure 310. DHE Database structure.

a wall colour of green on the node 15 side. (There will be a different colour attribute on the node 10 side.) The 'wall colour' field may, if necessary, provide a pointer to a texture map, allowing walk-through visualization.

Possible node attributes include a room number, possibly a volume, and even information about the escape route distance.

While the current implementation is graph-based, as objects in main memory, it is easy to store as four tables in a database (Figure 311).

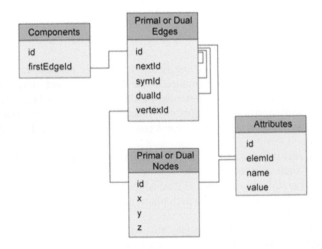

Figure 311. Database table structure.

Components: (usually one) Separate cell complexes not connected together, but being a part of the same collection of models.

Attributes: Attributes can be assigned to nodes and edges.

Nodes: (containing coordinates) Primal nodes represent nodes in the primal space and volumes in the dual space. Dual nodes represent nodes in the dual space and volumes in the primal space. There will thus be pointers to both primal and dual attributes.

Edges: Primal edges represent edges in primal space and faces in dual space. Dual edges represent edges in the dual space and faces in the primal space. There will thus be pointers to both primal and dual attributes.

Application

This data structure was applied to two connected buildings at the University of Glamorgan (now the University of South Wales). Floor plans were obtained, e.g. Figure 312, for each building and floor.

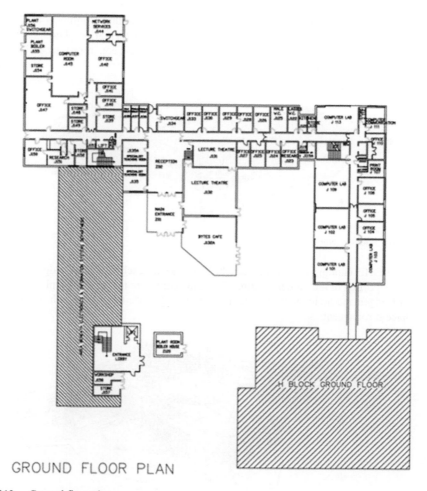

GROUND FLOOR PLAN

Figure 312.   Ground floor plan

AutoCAD was used to build a set of 3D boxes from each of these rooms, to a standard height (Figure 313). Note that these rooms are not connected – they merely sit next to each other.

199

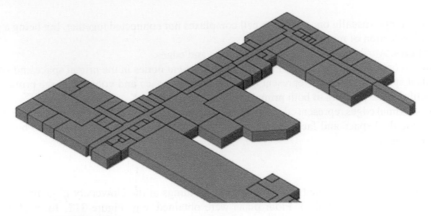

Figure 313. A single floor level.

The floors were then stacked up, along with a simple exterior surface (Figure 314). Doors were added as zero-thickness boxes inserted into the walls. Explicit representation of doors in the model allowed for a simple attribute attachment to the door dual nodes, e.g. room accessibility time, access card requirements, etc.

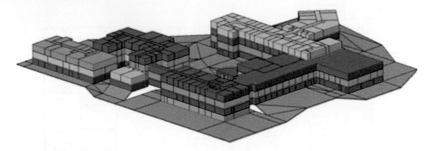

Figure 314. All floors together.

A complete DHE building model was then built, by taking each AutoCAD edge and adding it to the previous portion of the model (Figure 315). Blue edges are the primal graph (the geometry, composed of edges and nodes with coordinates). Red edges are the dual graph, connecting the various adjacent room centres.

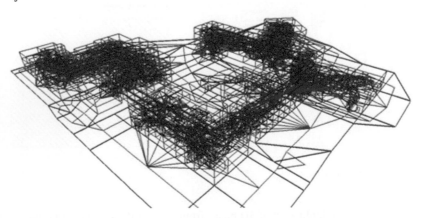

Figure 315. Complete DHE structure.

Here some of the room centres are shown (Figure 316), and represent the volumes as dual node balls, along with the primal nodes (room corners). Note the presence of doors, with their own volume centres.

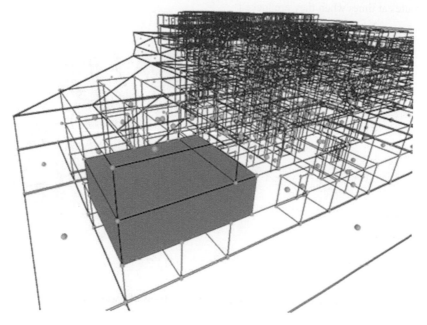

Figure 316.   Primal and dual nodes.

Standard graph traversal techniques, as described earlier, may be performed on the dual graph, showing a possible route from a particular room to the nearest exit (Figure 317).

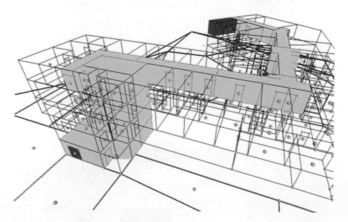

Figure 317.   Shortest path to an exit.

Doors are volume entities with connected dual edges to the adjacent rooms. The adjacent walls also have dual edges connecting the same pairs of rooms, but in order to restrict access through them an infinite weight is set, while in the case of connections through doors finite weights representing a distance are set. Of course, in some kinds of emergency escape scenarios, the difficulty of getting through the wall (by a fireman, perhaps) may be large but possible, and simulations of this type may be accommodated by adjusting the edge connection weight accordingly. Because the DHE may be locally updated, structural modification – perhaps a wall collapsing – may be incorporated into the simulation, as may time. Figure 318 shows the buildings' doors, any of which may have edge connection weights changed according to the time of day, for example, enforcing different escape routes at times when they are locked.

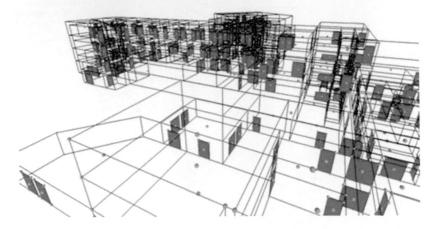

Figure 318.   Building doors.

Here emergency access (to the nearest washroom!) is straightforward (Figure 319).

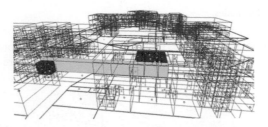

Figure 319.   Washroom access.

But in Figure 320 the connecting door (in red) is locked, enforcing travel to the washroom on the floor above.

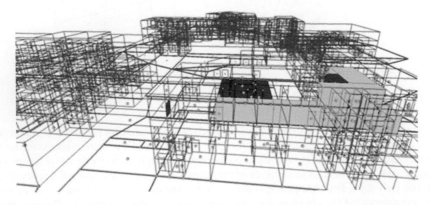

Figure 320.   After-hours access.

Depending on applications, different model representations may be taken into consideration. For example, it may be justified to add wall elements between adjacent rooms and represent them as volumes. This makes attachment of wall attributes to a wall dual node straightforward. Also, detailed structure and potential damage of walls in the case of a disaster may be explicitly reflected in the model, in real time.

3D Summary

In this section we have moved from the original 2D primal-dual Quad-Edge structure to a 3D model with the same features. Using the DHE makes 3D modelling (true 3D, with volume entities) straightforward, at much the same level of difficulty as 2D. Being entirely graph based, all the varied graph manipulation tools developed over the years within Computer Science may be applied, and various kinds of attributes used for a large selection of simulation problems.

Finally, it fits our early Cosmic Model, with voids and faces, providing a level of consistency to the modelling of Space in multiple disciplines.

FURTHER READING

Research in CAD systems led from the simple 2-manifold cases usually used, to the non-manifold cases needed where 3D cell complexes are required, as in building interiors. Important articles are those by Yamaguchi and Kimura (1995) and Lopes and Tavares (1997). The Dual Half-Edge is described in Boguslawski (2011), Boguslawski et al. (2011) and Boguslawski and Gold (2011). Molenaar (1998) gave a theoretical framework for 3D object modelling, Penninga (2008) described a system based on 3D simplicial complexes, and Kolbe et al. (2008) described the advantages of the CityGML building modelling standard for emergency response.

Boguslawski and Gold (2016a, 2016b) described using the Dual Half-Edge to integrate 3D building models and the exterior terrain, together with the detailed structure as used for building interior escape route planning. This class of application was described in Goodchild (2006) and Filippoupolitis (2009), and is becoming increasingly significant.

REFERENCES

Boguslawski, P. (2011). *Modelling and Analysing 3D Building Interiors with the Dual Half-Edge Data Structure*. PhD Thesis. University of Glamorgan.
Boguslawski, P. & Gold, C.M. (2011) Rapid modelling of complex building interiors. In: Kolbe, T.H., König, G. & Nagel, C. (eds.) *Advances in 3D Geo-Information Sciences. Lectures Notes in Geoinformation and Cartography*. Springer. pp. 43–56.
Boguslawski, P. & Gold, C.M. (2016a, in press) Buildings and terrain unified – Multidimensional dual data structure for GIS. *Geo-Spatial Information Science*, 18 (4), 151–158.
Boguslawski, P. & Gold, C.M. (2016b, in press) The dual half-edge – A topological primal/dual data structure and construction operators for modelling and manipulating cell complexes. *ISPRS International Journal of Geo-Information*.
Boguslawski, P., Gold, C.M. & Ledoux, H. (2011) Modelling and analysing 3D buildings with a primal/dual data structure. *ISPRS Journal of Photogrammetry and Remote Sensing – Theme Issue: Scale, Quality, and Analysis Aspects of City Models*, 66, 188–197.
Charlesworth, H.A.K., Langenberg, C.W. & Ramsden, J. (1975) Determining axes, axial planes and sections of macroscopic folds using computer-based methods. *Canadian Journal of Earth Sciences*, 13, 54–65.
Filippoupolitis, A., Loukas, G., Timotheou, S., Dimakis, N. & Gelenbe, E. 2009. Emergency response systems for disaster management in buildings. In: *Proceedings of the NATO Symposium on C3I for Crisis, Emergency and Consequence Management*. Romania, Bucharest, pp. 1–14.
Gold, C.M. (2005) Data structures for dynamic and multidimensional GIS. In: *Proceedings: 4th ISPRS Workshop on Dynamic and Multi-dimensional GIS*. pp. 36–41.
Goodchild, M.F. (2006) GIS and disasters: Planning for catastrophe. *Computers, Environment and Urban Systems*, 30, 227–229.

Kolbe, T.H., Gröger, G. & Plümer, L. (2008) CityGML – 3D city models and their potential for emergency response. In: Zlatanova, S. & Li, J. (eds.) *Geo-Information Technology for Emergency Response*. London, Taylor & Francis.

Ledoux, H. (2007) Computing the 3D Voronoi Diagram robustly: An easy explanation. In: *4th International Symposium on Voronoi Diagrams in Science and Engineering*. pp. 117–129.

Ledoux, H. & Gold, C.M. (2004) Modelling oceanographic data with the three-dimensional Voronoi Diagram. In: *Proceedings of the XXth ISPRS Congress*. Vol. 2. pp. 703–708.

Ledoux, H. & Gold, C.M. (2007). Simultaneous storage of primal and dual three dimensional subdivisions. *Computers, Environment and Urban Systems*, 31, 393–408.

Ledoux, H. & Gold, C.M. (2008) Modelling three-dimensional geoscientific fields with the Voronoi Diagram and its dual. *International Journal of Geographical Information Science*, 22, 547–574.

Lee, K. (1999) *Principles of CAD/CAM/CAE Systems*. Boston, MA, Addison-Wesley Longman Publishing Co., Inc. 582 pp.

Liu, Y. & Snoeyink, J. (2005) A comparison of five implementations of 3D Delaunay tessellation. *Combinatorial and Computational Geometry*, 52, 439–458.

Lopes, H. & Tavares, G. (1997) Structural operators for modeling 3-manifolds. In: *Proceedings, 4th ACM Symposium on Solid Modeling and Applications*. pp. 10–18.

Maas, H.-G. & Vosselman, G. (1999) Two algorithms for extracting building models from raw laser altimetry data. *ISPRS Journal of Photogrammetry & Remote Sensing*, 54, 153–163.

Mäntylä, M. (1986) Boolean operations of 2-manifolds through vertex neighbourhood classification. *ACM Transactions on Graphics*, 5, 1–29.

Mäntylä, M. (1988) *Introduction to Solid Modeling*. New York, NY, Computer Science Press, Inc. 401 pp.

Maxwell, J.C. (1864) On reciprocal diagrams and diagrams of forces. *Philosophical Magazine*, 4, 250–261.

Molenaar, M. (1998) *An Introduction to the Theory of Spatial Object Modeling for GIS*. London, Taylor & Francis.

Penninga, F. (2008) *3D Topography: A Simplicial Complex-Based Solution in a Spatial DBMS*. PhD Thesis. Delft, Delft University of Technology. 192 pp.

Sazanov, I., Hassan, O., Morgan, K. & Weatherill, N.P. (2007) Generating the Voronoi–Delaunay dual diagram for co-volume integration schemes. In: *The 4th International Symposium on Voronoi Diagrams in Science and Engineering (ISVD 2007)*. pp. 99–204.

Sugihara, K. & Hiroshi, I. (1995) Why is the 3D Delaunay triangulation difficult to construct? *Information Processing Letters*, 54, 275–280.

Tse, R.O.C. & Gold, C.M. (2002) A surface-representation approach to a three-dimensional cadastre. *The GIM International Journal*, 16, 46–49.

Tse, R.O.C. & Gold, C.M. (2003) A proposed connectivity-based model for a 3D cadastre. *Computers, Environment and Urban Systems*, 27, 427–445.

Tse, R.O.C. & Gold, C.M. (2004) TIN meets CAD – Extending the TIN concept in GIS. *Future Generation Computer Systems (Geocomputation)*, 20, 1171–1184.

Tse, R.O.C., Gold, C.M. & Kidner, D. (2007) Using Delaunay triangulation/Voronoi Diagram to extract building information from raw LIDAR data. In: *4th International Symposium on Voronoi Diagrams in Science and Engineering*. pp. 222–229.

Tse, R.O.C., Gold, C.M. & Kidner, D. (2008a) 3D city modelling from LIDAR data. In: van Oosterom, P., Zlatanova, S., Penninga, F. & Fendel, E. (eds.) *Advances in 3D Geoinformation Systems*. Springer-Verlag Berlin Heidelberg. pp. 161–175.

Tse, R.O.C., Gold, C.M. & Kidner, D. (2008b) Implementation of building reconstruction algorithm using real world LIDAR data. In: *SDH-SAGEO 2008 Conferences*. pp. 297–313.

Yamaguchi, Y. & Kimura, F. (1995) Non-manifold topology based on coupling entities. *IEEE Computer Graphics and Applications*, 15, 42–50.

Chapter 7

Conclusions

7.1   WHAT HAVE WE LEARNED?

We have shown how vectors get us off to a good start in freeing ourselves from the complexities of coordinate systems – and how a couple of simple predicates facilitate the geometric operations greatly. Not only do they help us with basic area, volume and intersection testing, but the concept of sidedness, in 2D and 3D, is fundamental to obtaining well-posed questions about the spatial relationships of lines and points, leading us to the understanding of 'good' coordinate systems – local ones, adaptive to the distribution of the data, and not based on some imaginary grid system in the sky. If you gain nothing more than an understanding of simple barycentric coordinates and the CCW test, this book will have been worthwhile.

This leads naturally to the concept of dominance and the Voronoi diagram – an apparently 'natural' concept discovered many times in different ages and disciplines, and one that should now pop up in your examination of many aspects of the natural world – from tree growth to giraffe skins to frying eggs in a pan. And by this time you know enough to write your very own Voronoi code – because now, reluctantly, we have to come back to computers and how to stuff space into them. All is not lost, however, as they happen to be very good at handling graphs – connected collections of edges and points – how to navigate in them, and the fundamental concept of duality, which gives us the key clue about the spatial 'context' of an object, as well as some ideas about data structures: how to represent these graphs so they can be stored handily in a computer's memory or disc.

We then go on to puzzle over the basic question of what is space, anyway – so, like the early scientists, we look at the universe to get some clues. In our case, because of modern cosmology, we can see a possible picture of galaxies and force fields – giving us the concept of boundaries and voids as our underlying elements. This handily fits with our Voronoi model. It also leads to a variety of boundary types, which can be matched to appropriate computer data structures in 2D and 3D.

The first type of boundary is – none at all, just individual points. Here the VD/DT provides the context for each object. This can be used for nearest-point interpolation, or for clumping them together if the points are labelled. This gives us a skeleton or boundary around each group of similar points. It also forms the basis of the simple TIN terrain model, where the points are linked together by the triangles. A more advanced interpolation that avoids the more traditional 'counting circle' methods is the 'area-stealing' approach, where query points are added and then removed, and the neighbouring Voronoi areas eaten by the new point's cell are used for a weighted average. This has many advantages for precise data, especially if the points are anisotropically distributed: a 'reasonable' set of neighbouring data points is produced, and the surface slopes are guaranteed to be continuous except at the data points themselves.

A rather interesting addition is the kinetic algorithm, which handles moving points and their spatial context, and can be used for flow modelling or collision-detection systems. This takes us beyond static maps, and into the realm of simulation and decision support. But first we must look further at boundaries.

The next type is based on a single string of points – but we must be able to identify that it is a string, or a 'crust', based on Blum's concept of a 'skeleton' mid-way between two portions of a boundary, and Amenta's definition, followed by Gold's circumcircle test to separate VD/DT edge pairs into either crust or skeleton segments. This, along with basic smoothing operations, gives well-defined object boundaries (crusts) as well as interior and exterior skeletons, permitting the clear identification of polygonal objects and their separators or interior structures (skeletons). In a frequent geographic context this is usefully applied to digitized contour lines, to identify ridge or valley lines and estimate some additional intermediate elevations.

Another way of looking at terrain is to construct the crust and skeleton network based on the hydrography – the crust guarantees river continuity and the skeleton approximates the various levels of watershed, mid-way between the river segments, using Blum's height transform to estimate ridge heights. Examining the Voronoi cells of each reach of the river gives their 'micro-catchment' areas, and summing these then models a static accumulation of rainfall runoff within the system, useful for preliminary flooding analysis.

However, in practice the most important aspect of terrain interpolation is not necessarily precise elevation estimates, as not many geographic processes are dependent on exact heights, but the surface slope, which is critical for vegetation patterns, etc. – and runoff, as just described for modelling based on the hydrography. So it is important to incorporate slopes into traditional interpolation: any reasonable slope estimate at the data point will do, as long as it is used in the local weighted average calculation. Combining data point slope estimates with an adaptive neighbour selection process produces the best surface slope models so far – using either Sibson interpolation or planar interpolation within each triangle. We illustrate this with a Voronoi-based runoff simulation, which should show up any computational artefacts in the method.

We then move on to other boundary types, based on our Cosmic Model. As the galaxies move, under the influence of gravity, the voids tend to grow and boundaries form between them. These 'slabs' may be several galaxies thick, with the gravitational zones of influence of the exterior ones filling the voids on each side, forming what we think of as polyhedra or polygons. These exterior points form a chain (in 2D) defining the extent of the empty voids on each side: the interior points, in this model, have less influence but form a type of cluster, with multiple holes, and various types of cluster analysis may be used – in particular, for our purposes, analysis of the Voronoi cell shape to determine cluster boundaries (the same thing as void boundaries, if you think about it – just looking at them from the other side). A particularly useful form of cluster analysis uses the Euclidean Minimum Spanning Tree, a subset of the Delaunay triangulation – based on the connectivity of adjacent points: the dual is the context, again.

If we remove the points interior to the slab – the filling in the sandwich – we end up with the double-point boundary: two crusts with an intermediate skeleton. This has been found useful in various types of mapping problems – forest polygon digitizing in our case: the skeleton forms the boundary between the polygons. Its particular strength is that it has guaranteed topological connectivity, as we mentioned in the case of labelled skeletons, and it can be used to reconstruct polygon boundaries from interior polygon points. (In practice another step is needed – a scanning sequence for the triangles, to ensure that the individual skeleton segments are connected up efficiently.) The same process can be used for scanned maps, such as cadastral plans, if an edge detection filter is used to detect the boundaries between 'black' and 'white' pixels. The resulting skeletons may be used for recognising the labels interior to individual properties, as well as determining the topologically connected property boundaries. (For very large data sets a hierarchical indexing system is needed for rapid searching or for paging map segments onto disc, or for parallel processing. This can be achieved by indexing lower level Voronoi generators within the Voronoi cells of the higher levels.)

A final boundary type for our geographic applications is found when we replace the previous 'sandwich' by a solid line along the skeleton path. This requires a new type of Voronoi generator, a line segment. While extremely valuable in practice, it is unfortunately extremely complex to implement, due to the large number of special cases occurring when lines meet or intersect. A simpler approximation, the Constrained DT, forces particular triangle edges to conform to some geographic feature – perhaps a building edge or a river segment – but there are other types of special cases and these edges, which should be map entities like the vertices, are confused with the DT, which should be concerned exclusively with the spatial context of the entities. While various algorithms have been proposed, here we are taking the mathematical statement that a line or curve is the locus of a moving point as the basis for a methodology founded on the kinetic moving point described previously.

If we wish to rigorously separate the map entities and their context, we need to create half-line map entities, themselves fully connected by the dual DT. In particular the circumcircle tangent to three line segments or points requires careful calculation. We have again taken the kinetic moving point as our starting point, as it permits interactive collision detection and intersection, as well as defining the map construction operation as a sequence of commands that may be rolled forward or rolled back as required, although the method of Martin Held is probably the most robust to date. Mioc (2012 etc.) has developed the logic of dynamic map modification, leading to a temporal tree structure for recovering different map versions.

Having separated map entities from their context permits various simplified basic queries, such as 'What is here?' (more properly, 'What is closest to here?') and buffer zones (which are merely clipped Voronoi regions). Initially partitioning the map into proximal regions greatly simplifies subsequent queries. As the line segments are really half-lines associated with a particular side, and likewise for their associated Voronoi cells, map partitioning may be achieved in a hierarchical manner if desired, removing polygon interiors for storage elsewhere.

With our basic toolkit we are ready to design our 2D Spatial Decision Support System – a GIS designed for interactive query and modification, as well as various types of simulation, designed from the ground up for rapid editing and interaction, assisted by the unified Voronoi spatial model. We then move on to higher dimensions – and the embedding of lower dimension spatial structures within them.

What we have learned about Quad-Edges also applies to CAD (Computer Aided Design) systems: a small extension gives us the Euler Operators we need for incremental construction of 'Shells' – closed boundary surfaces of arbitrary objects, such as buildings, which may be constructed independently or extruded from the landscape. Another minor adaptation allows for the construction of bridges and tunnels: if these are extruded from the landscape flood modelling or similar operations may take account of these features.

However, frequently we do not have this design information, and must rely on modern LiDAR elevation data, which must be converted from point clouds on the building surface into CAD building faces. If we do not have appropriate building templates, this requires a series of carefully defined propositions to help identify and construct the building: the existence of a high region entirely surrounded by vertical walls; roofs composed of triangular plates whose orientations cluster together, etc. From these the wall and roof planes can be estimated, and their intersections and the building faces determined. In this approach an extension of the 'Polyhedral Earth' is constructed – but one with no interior volumes, merely the boundary between earth and air.

When we wish to distinguish between different volumes – different rooms, for example, or rock types – we must include a 'Volume' entity along with the traditional 'Faces', 'Edges' and 'Nodes'. For the VD/DT these cells may be constructed using operations similar to the 2D flipping, and the Delaunay vertices would contain the attributes for the associated Voronoi cell. Simple contour surfaces may be constructed by slicing the Delaunay triangles appropriately. This gives a useful way to contour 3D data such as ships' oceanographic data, e.g. temperature and salinity – this data is collected in a way that is anisotropic in Z (depth) as well as in X and Y (the ships' tracks) and, as in 2D, the data structure is adaptive to the real data density in any direction. Full 3D Sibson interpolation may also be implemented if required.

For more general 3D structures, such as buildings with rooms, a new data structure was required that incorporated the volume elements. The Augmented Quad-Edge was developed to allow navigation between the b-rep surfaces of each Voronoi cell, and was extended as the Dual Half-Edge to permit CAD-type Euler construction operators, that maintain complete navigability in both the primal and dual graphs while individual edges are being added or removed. This means that navigation from room to room, along the dual edges between volume nodes, can be implemented to facilitate escape route planning and simulation – and to take account of any real-time changes in the building structure. Here, as in 2D, the dual graph provides the context, or relationships, for each room. Thus a consistent approach has been developed for geographical modelling in both 2D and 3D environments.

In summary, following Einstein (1952):

> "I wished to show that space-time is not necessarily something to which one can ascribe a separate existence, independently of the actual objects of physical reality. Physical objects are not in space, but these objects are spatially extended. In this way the concept of 'empty space' loses its meaning."

On this basis, only the inclusion of a dual representation, in any dimension, permits the existence of a full set of ontological elements – areas in 2D, volumes in 3D. Without these the space is incomplete, queries become complex and perverse, and the representation of spatial reality becomes increasingly implausible.

2D and 3D entities may be subdivided, connected, queried and stuffed in the computer. The primal entity is a fragment of the picture – of the jigsaw. The dual entity is the interlocking boundary that helps construct the whole.

The spatial dual is the context.

Index

Page numbers in **bold** indicate figures.

ISPRS Book Series

1. Advances in Spatial Analysis and Decision Making (2004)
   Edited by Z. Li, Q. Zhou & W. Kainz
   ISBN: 978-90-5809-652-4 (HB)

2. Post-Launch Calibration of Satellite Sensors (2004)
   Stanley A. Morain & Amelia M. Budge
   ISBN: 978-90-5809-693-7 (HB)

3. Next Generation Geospatial Information: From Digital Image Analysis to Spatiotemporal
   Databases (2005)
   Peggy Agouris & Arie Croituru
   ISBN: 978-0-415-38049-2 (HB)

4. Advances in Mobile Mapping Technology (2007)
   Edited by C. Vincent Tao & Jonathan Li
   ISBN: 978-0-415-42723-4 (HB)
   ISBN: 978-0-203-96187-2 (E-book)

5. Advances in Spatio-Temporal Analysis (2007)
   Edited by Xinming Tang, Yaolin Liu, Jixian Zhang & Wolfgang Kainz
   ISBN: 978-0-415-40630-7 (HB)
   ISBN: 978-0-203-93755-6 (E-book)

6. Geospatial Information Technology for Emergency Response (2008)
   Edited by Sisi Zlatanova & Jonathan Li
   ISBN: 978-0-415-42247-5 (HB)
   ISBN: 978-0-203-92881-3 (E-book)

7. Advances in Photogrammetry, Remote Sensing and Spatial Information Science. Congress
   Book of the XXI Congress of the International Society for Photogrammetry and Remote
   Sensing, Beijing, China, 3–11 July 2008 (2008)
   Edited by Zhilin Li, Jun Chen & Manos Baltsavias
   ISBN: 978-0-415-47805-2 (HB)
   ISBN: 978-0-203-88844-5 (E-book)

8. Recent Advances in Remote Sensing and Geoinformation Processing for Land Degradation
   Assessment (2009)
   Edited by Achim Röder & Joachim Hill
   ISBN: 978-0-415-39769-8 (HB)
   ISBN: 978-0-203-87544-5 (E-book)

9. Advances in Web-based GIS, Mapping Services and Applications (2011)
   Edited by Songnian Li, Suzana Dragicevic & Bert Veenendaal
   ISBN: 978-0-415-80483-7 (HB)
   ISBN: 978-0-203-80566-4 (E-book)

10. Advances in Geo-Spatial Information Science (2012)
    Edited by Wenzhong Shi, Michael F. Goodchild, Brian Lees & Yee Leung
    ISBN: 978-0-415-62093-2 (HB)
    ISBN: 978-0-203-12578-6 (E-book)

11. Environmental Tracking for Public Health Surveillance (2012)
    Edited by Stanley A. Morain & Amelia M. Budge
    ISBN: 978-0-415-58471-5 (HB)
    ISBN: 978-0-203-09327-6 (E-book)

12. Spatial Context: An Introduction to Fundamental Computer Algorithms for Spatial Analysis (2016)
    Christopher M. Gold
    ISBN: 978-1-138-02963-7 (HB)
    ISBN: 978-1-4987-7910-4 (E-book)

Printed and bound by CPI Group (UK) Ltd, Croydon, CR0 4YY
01/11/2024
01782604-0001